Principles of Ecology

RICHARD BREWER

Western Michigan University
Kalamazoo

1979   W. B. SAUNDERS COMPANY / Philadelphia / London / Toronto

W. B. Saunders Company: West Washington Square
Philadelphia, PA 19105

1 St. Anne's Road
Eastbourne, East Sussex BN21 3UN, England

1 Goldthorne Avenue
Toronto, Ontario M8Z 5T9, Canada

Library of Congress Cataloging in Publication Data
Brewer, Richard.
    Principles of ecology.
    Includes index.
    1. Ecology. I. Title.
QH541.B73 1979   574.5   77-84666
ISBN 0-7216-1988-6

Cover photograph of the prairie forb wild indigo (*Baptisia leucantha*) by the author.

Principles of Ecology   ISBN   0-7216-1988-6

Last digit is the print number: 9 8 7 6 5 4 3 2 1

PREFACE

There are many good ecology textbooks but most of them are aimed at the upper undergraduate or graduate student. This is an undergraduate text designed for a sophomore–junior level course in general ecology. It assumes no knowledge beyond that usually attained by the time freshman biology is completed.

The language has nearly always been kept as plain as I can write. Special ecological terms are defined where they first occur, either directly or by context, as are special physiological, chemical, and other scientific terms.

Material within the six chapters is arranged in a fairly large number of rather brief sections. One of the considerations in the placement of material into sections was to allow some of the more complex material to be omitted without losing the fundamentals of the subject. For example, there are two sections on the ecological niche, one a simple descriptive treatment and one which develops G. E. Hutchinson's hypervolume model.

Most of the text is not mathematical but in a few places sections introducing certain mathematical aspects of ecology have been included. These sections require some knowledge of algebra, and a little calculus would help for some of them. They present what I hope are thorough, step-by-step explanations that even poor mathematicians may follow. For some introductory ecology courses these mathematical sections may be unnecessary and may be omitted with no loss of continuity.

I have taken particular care with the figures and tables, which are used to clarify or supplement the text. In several cases they extend the text by giving a more detailed treatment of a particular topic. For courses in which such detail is unnecessary the instructor can readily designate any tables or figures he does not wish to assign. Because beginning students sometimes have trouble interpreting graphs and tables, I have tried to provide fully descriptive captions.

The bibliographies at the end of each chapter are selective. They were designed to permit further study of a subject by including some of the more recent reviews in article or book form, some of the classic papers on the subject, and some recent research papers.

To save space and to avoid the dampening effect that scholarly apparatus has on some students, I have not included scientific names or in-text citations of literature. These omissions will cause few problems for the intended audience. Most common names used are standard ones, which can be readily connected with the appropriate Latin names in standard taxonomic sources.

This is a book of ecology; its framework is the major principles of the science. I have tried to cover these at all levels, from the ecology of the individual—a neglected area in most current textbooks—to that of the ecosystem. Many applied or practical aspects of ecology are discussed in the text in connection with appropriate ecological principles, with the final chapter dealing with the role of ecology in human affairs. Clearly, the book gives no complete treatment of the topic of the environment and man, but I believe that the treatment it does give is valuable on two counts. First, most of the unifying principles of environmental science are ecological. The names of pesticides may change but the rules that organisms, populations, and ecosystems follow do not. Accordingly, a knowledge of the basic principles of ecology and physiology prepares one as well for understanding pollution or the population problem as do catalogs of food additives, herbicides, or contraceptive devices.

I do not, however, mean to make light of the latter kind of specific information on environmental matters. My second reason for believing that human environmental concerns should be included in a basic ecology text is this: I see entirely too few biology majors and minors who have any sound knowledge of environmental problems. Too often it is assumed that the student will have learned of these problems by reading the newspaper and will have made the connection with the appropriate ecological or biological principles on his own, and too often this has not happened. Biologists must be ready to take a part in the solving of environmental problems—they are too important to be left to social scientists and engineers.

I have learned much of what I know about ecology and environmental matters directly from other individuals. I am grateful, first, to my formal teachers of ecology: S. C. Kendeigh, W. L. Gersbacher, the late A. G. Vestal, and J. W. Voigt. I am grateful to colleagues from boyhood to the present: J. W. Hardy, K. D. Stewart, C. Heckrotte, W. B. Robertson, Jr., G. C. West, G. W. Cox, W. L. Gillespie, T. S. Robinson, W. J. Davis, W. L. Minckley, R. W. Olsen, C. G. Goodnight, R. W. Pippen, J. G. Engemann, A. M. Laessle, D. L. Regehr, and M. L. Kaufmann. I am grateful, also, to the students who, in class and out, have tried to educate me. There are too many to list but I should mention M. T. McCann, who read and commented on the whole book.

Western Michigan University has provided support over the years.

Finally, for many favors, I am grateful to my wife, Lucy Sharp Brewer, to whom this book is dedicated, of course.

RICHARD BREWER

CONTENTS

Chapter 1

ECOLOGY AS A SCIENCE

Ecology is a relatively new science. The word was first defined just over a hundred years ago by the German zoologist Ernst Haeckel. He based it on the Greek word *oikos* meaning home, and wrote: "By ecology we mean the body of knowledge concerning the economy of nature—the investigation of the total relations of the animal both to its inorganic and its organic environment." This is basically the same definition as the one used today—the study of the relationships of organisms to their environment and to one another.

Ecology has gone from a word that few people knew a decade ago to one that is widely misused today (Fig. 1–1). A popular entertainer has been quoted as saying that he was traveling around the country preaching ecology. (If this statement does not sound a little odd, try substituting the name of some other science, say anatomy, for ecology.) Environmental concern is probably what is being preached; it is not ecology but is an activity to which ecology has a great deal to contribute. The phrase "environmental science" covers all the sciences—including ecology, geology, and climatology—that deal with the environment. A good term for these and all the other fields that have an interest in the use of the environment might be "environmental studies"; a list of these would run from economics to religion and back again.

Although ecology had a name by 1869, there was at that time no "body of knowledge" because almost no one was engaged in the kind of research that yielded ecological information. More and more such studies were begun in the later years of the nineteenth century, but it was not until about 1900 that certain biologists began to think of themselves as ecologists and the kind of work they were doing as ecology.

Although Germany, Denmark, France and other countries had scientists who contributed to the early development of the field of ecology, a good share of the early work at the beginning of this century was done in the United States. We can imagine an early ecologist setting out with

1

"How can a guy as old as you be a professor of ecology
when it didn't even <u>exist</u> six or seven years ago?"

Figure 1–1. Few ecologists would agree that Montgomery Ward's is anybody's ecology headquarters, but few will object to the public's casual use of the term as long as it indicates a knowledge of the earth's environmental problems and support for efforts to solve them. The cartoon (from *Audubon, 79*(3):121, 1977) illustrates a common misconception.

Figure 1–2. A field trip 20 May 1913 led by the pioneer ecologist C. C. Adams from the University of Illinois (Western Michigan Univ. Archives).

a class from the University of Illinois or Chicago or Nebraska (Fig. 1–2). The instructor and the students wear dark suits, ties, and hats after the fashion of the period. They travel by electric railroad to the station nearest their study area, and then walk. The study area may be a beech-maple forest, the beaches and dunes of Lake Michigan, or a bog. They conduct their studies, eat the sandwiches their landladies have fixed for them, and at the end of the day walk back to the station for the trip home. If some of the students are planning to become ecologists, they might be needled a bit by their classmates. What difference does it make what plants grow on sand dunes? What can a study of the animals that live on the bottom of the Illinois River tell us? Why waste time studying ecology when you can work on a good solid topic like the embryology of the sea urchin or go to medical school?

Times change. If we were to visit the bog studied by that class of 60 or 70 years ago we would travel by expressway rather than by interurban, and we might find, not lady's slipper orchids and cranberries, but a garbage dump or a housing development built on one. The beach would be different, too; there might be beer cans instead of sea rockets, and there would be other changes, some visible, some not. There would be large numbers of people, tanned and pale, slim and fat, packed onto the beach at a density greater than in the apartment houses they live in. Some chemical changes in the lake water and the organisms in it might not be visible; we cannot see the DDT in the fat of the fishes or the increased phosphate content of the water directly.

Times change, and the science of ecology is no longer something of interest only to ecologists or a subject to be studied only for its intellectual fascination (although that still exists). Many of the major problems the world faces—pollution, overpopulation, the wise use of resources—are at heart ecological problems. Ecologists will help to find solutions to these problems, but so will everyone else. A knowledge of environmental science is necessary to all persons so that they may live, conduct their business, and vote in such a way so as to make solutions possible.

In the following chapters we will examine the science of ecology at three levels. The ecology of the *individual organism* will be considered first. The second level deals with the ecology of groups of individuals, or *populations*. Populations of several kinds of organisms, several species, live together in *communities*, the third level. At this same level we may consider the community along with its physical setting or habitat as a single, interacting unit, the **ecosystem.** A pond is a familiar ecosystem, with the plants and animals and bacteria forming the community, and the water, the dissolved salts and gases, and the mud of the bottom being elements of the rest of the system. We can recognize communities and ecosystems of various scales, from the stomach of a cow, with its interesting populations of microorganisms, on up to the earth itself, the largest ecosystem with which we are familiar.*

One of the great lessons of ecology is the interrelatedness of nature, so there is some justification for thinking of the ecosystem as the fundamental unit of study. There is a story that a few years ago DDT was used in Borneo to kill mosquitoes around houses, which it did. However, small lizards called geckos who lived in the houses and fed on insects began to die from their DDT-rich diet. Weakened geckos fell easy prey to house cats who also began to die. As the cat population dropped rats began to infest the houses, and in this area rats were dangerous as potential plague carriers. Borneo began to import cats.

I do not know whether this story is strictly true; I have not seen it documented. Whether or not it is true it expresses clearly the spirit of the profound truth that the parts of an ecosystem are all interconnected so that when we touch one part of the system, we eventually and in some way touch the rest.

BIBLIOGRAPHY: TEXTBOOKS IN ECOLOGY

The field of ecology is fortunate in having a great many excellent textbooks. Because books in the following list have valuable information and ideas on a wide variety of topics, they are brought together here

---

*The entire global environment supporting life from the depths of the oceans and as far down in the soil as organisms occur up to the highest part of the atmosphere occupied by organisms is termed the *biosphere.*

rather than being mentioned repeatedly in the bibliographies of later chapters.

Allee, W. C., Emerson, A. E., Park, O., Park, T., and Schmidt, K. P. *Principles of Animal Ecology.* Philadelphia, W. B. Saunders, 1949.

Andrewartha, H. G., and Birch, L. C. *The Distribution and Abundance of Animals.* Chicago, University of Chicago Press, 1954.

Bodenheimer, F. S. *Animal Ecology Today.* The Hague, Dr. W. Junk, 1958.

Chapman, R. N. *Animal Ecology with Especial Reference to Insects.* New York, McGraw-Hill, 1931.

Clarke, George L. *Elements of Ecology.* New York, John Wiley & Sons, 1954.

Clements, Frederic E., and Shelford, V. E. *Bio-Ecology.* New York, John Wiley & Sons, 1939.

Colinvaux, P. A. *Introduction to Ecology.* New York, John Wiley & Sons, 1973.

Collier, B. D., Cox, G. W., Johnson, A. W., and Miller, P. C. *Dynamic Ecology.* Englewood Cliffs, N.J., Prentice-Hall, 1973.

Daubenmire, R. F. *Plant Communities: A Textbook of Plant Synecology.* New York, Harper & Row, 1968.

————. *Plants and Environment,* 3rd ed. New York, John Wiley & Sons, 1974.

Emlen, J. M. *Ecology: An Evolutionary Approach.* Reading, MA, Addison-Wesley, 1973.

Hanson, H. C., and Churchill, E. D. *The Plant Community.* New York, Reinhold, 1961.

Kendeigh, S. Charles. *Ecology with Special Reference to Animals and Man.* Englewood Cliffs, N.J., Prentice-Hall, 1974.

Kershaw, K. A. *Quantitative and Dynamic Ecology.* New York, American Elsevier, 1973.

Kormondy, E. J. *Concepts of Ecology,* 2nd ed. Englewood Cliffs, N.J., Prentice-Hall, 1976.

Krebs, C. J. *Ecology: The Experimental Analysis of Distribution and Abundance.* New York, Harper & Row, 1972.

MacArthur, R. H. *Geographical Ecology.* New York, Harper & Row, 1972.

MacFadyen, A. *Animal Ecology: Aims and Methods,* 2nd ed. London, Pitman, 1963.

May, R. M., ed. *Theoretical Ecology: Principles and Applications.* Philadelphia, W. B. Saunders, 1976.

McDougall, W. B. *Plant Ecology,* 3rd ed. Philadelphia, Lea & Febiger, 1941.

Odum, E. P. *Fundamentals of Ecology,* 3rd ed. Philadelphia, W. B. Saunders, 1971.

Oosting, H. J. *The Study of Plant Communities,* 2nd ed. San Francisco, Freeman, 1956.

Pianka, E. R. *Evolutionary Ecology.* New York, Harper & Row, 1974.

Pielou, E. C. *An Introduction to Mathematical Ecology,* 2nd ed. New York, John Wiley & Sons (Interscience), 1977.

Poole, R. W. *An Introduction to Quantitative Ecology.* New York, McGraw-Hill, 1974.

Ricklefs, R. E. *Ecology.* Portland, OR, Chiron, 1973.

————. *The Economy of Nature.* Portland, OR, Chiron, 1976.

Shelford, V. E. *Laboratory and Field Ecology.* Baltimore, Williams & Wilkins, 1929.

Smith, R. L. *Ecology and Field Biology,* 2nd ed. New York, Harper & Row, 1974.

Weaver, J. E., and Clements, F. E. *Plant Ecology.* New York, McGraw-Hill, 1938.

Whittaker, R. H. *Communities and Ecosystems.* New York, Macmillan, 1975.

Chapter 2

ECOLOGY OF INDIVIDUAL ORGANISMS

TOLERANCE RANGE

We refer to the surroundings of an organism as its **environment** or **habitat.** The basic subdivision is between *terrestrial*, or land, habitats and *aquatic*, or watery, habitats. A trout is an aquatic organism, living in cool streams, while a trout lily is terrestrial, living in moist forests. Aquatic habitats are often subdivided into *marine*, or ocean, habitats, and *freshwater* habitats which are lakes, ponds, streams, and springs whose waters have low salt concentrations. Much finer subdivisions can be made—habitats may be subdivided into subhabitats and eventually *microhabitats*, such as a south-facing slope, a tree hole, or the spaces between the sand grains along a beach.

Specific features of the habitat may be studied. For the trout we can measure such factors as water temperature and the amount of oxygen dissolved in the water. For the trout lily we can measure the amount of light reaching it on the forest floor and the concentration of calcium in the soil. If we are able to determine the whole range over which the species is able to live, for one of these factors, we then will know the **range of tolerance** of that species for that factor. A goldfish, for example, might be able to live in waters in which the temperature ranges from 2 to 34°C (about 36 to 93°F), but it will die of heat or cold at temperatures above or below this range.

Field observations give us some idea of the range of tolerance of a species, but the field work must be checked by experimentation. One of the reasons for this is that organisms interact with other organisms. The habitat features we have mentioned so far have all been *abiotic factors*, or factors of the physical environment. The organism also exists in a world

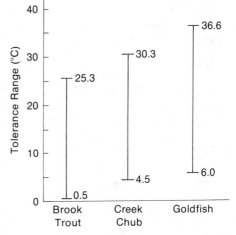

Figure 2–1 Different species have different ranges of tolerance. The goldfish is able to live in warmer waters than the brook trout but does not tolerate cold temperatures as well as the brook trout. (From J. R. Brett, "Some principles in the thermal requirements of fishes," *Quarterly Review of Biology,* 31:75, 1956.)

of *biotic factors,* consisting of its relationships with other organisms. These relationships are of several kinds: one organism may use another as food and at the same time serve as food for a third; two organisms may compete for food, for nesting sites, or for some other requirement. An organism may not be able to live in an area where physical factors are favorable for it due to some biotic factor. A common weed called sheep sorrel provides a good example. If we test the soil where these plants naturally grow, we almost always find that the soil is acid. If we plant its seeds and grow the plant in a greenhouse in soils of various acidity and alkalinity, we find that it grows best in neutral or slightly alkaline soils. Why is there this inconsistency? The answer seems to be that many plant species can grow in neutral soil and some of these can outcompete sheep sorrel under such conditions. Relatively few plants, however, can grow well in strongly acid soil. Thus, in nature, sheep sorrel is restricted by competition to one end of its tolerance range.

Tolerance ranges differ from one species to another. We would expect carp or goldfish, for example, to be able to tolerate warmer temperatures and lower oxygen concentrations than trout, and this is exactly what we find when we make the appropriate tests (Fig. 2–1). The fact that different species have different tolerance ranges is one basic reason why different habitats support different communities, and why communities vary geographically from warmer to cooler and from wetter to drier regions.

WHEN CONDITIONS CHANGE

Few habitats stay the same for very long. Some, such as the ocean depths, come very close, but for most habitats environmental factors change between day and night, between drought and a wet period, between summer and winter. When some important feature of the habitat changes, the organism changes in response. The change may be mainly *physiological*. As it warms up during the day, a plant and a bird both exposed to the sun will undergo physiological changes because of the heat. These changes often have the effect of keeping certain important aspects of the organism's internal environment constant despite the changing external environment. This tendency, which is an important physiological principle, is known as *homeostasis*. The bird may respond by lowering its rate of heat production and by arranging its feathers so as to lose heat rapidly, resulting in its body temperature remaining constant. If the sun keeps shining and the temperature keeps rising, an important difference between the plant and the bird or the trout lily and the trout may be seen. The plants are stationary, attached to one spot (*sessile*), whereas the animals can move about (*motile*). The bird or trout can move away from unfavorable conditions—the bird can go into the shade, the trout can swim to cooler water. These are *behavioral* changes which occur in addition to the purely physiological ones. The range of behavioral responses for a sessile organism is much smaller. Usually it can only remain in one place and tolerate the changed condition—or, if the change is too great, fail to tolerate it.

ACCLIMATION

There is one kind of physiological change shown by organisms that deserves special attention. The range of temperature tolerance of a fish (or an insect or a pine tree) may vary from season to season, depending on the temperature at which it has recently been living (Fig. 2–2). In

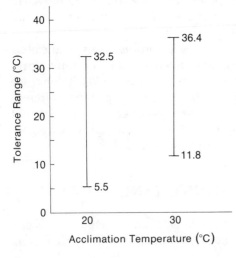

Figure 2–2. The tolerance range of organisms may be changed by acclimation. Kept at 20°C, the large-mouthed bass has an upper limit of temperature tolerance of just over 32°C. If the aquarium temperature is slowly raised to 30°C and kept there for a few days, the fish now acclimated to 30°C are able to tolerate temperatures of over 36°C. At the same time, however, they lose some of their ability to tolerate lower temperatures. (From Brett.)

midwinter, a fish may be able to live in the range from 0 to 24°C but in summer the same individual may have a tolerance range that extends all the way up to 33°C but down only to 15°C. Such adjustments in the ecological response to a changed environment are known as *acclimation.* Acclimation does not occur just in temperature tolerance; organisms may also acclimate to different oxygen levels in the air (if they go to higher altitudes, for example) and to other changed conditions.

Acclimation is undoubtedly of great importance in allowing organisms to exist permanently in changeable environments. Without this ability many organisms would either die or be forced to migrate during unfavorable seasons. Of course, some organisms do show these two strategies. Many birds, which are probably the most highly mobile of organisms, move back and forth seasonally between a summer and a winter range. Such organisms as annual weeds and many insects die each winter but produce seeds or eggs that have a much lower temperature tolerance limit than the parent organism.

ECOTYPES

The fact that two individuals belong to the same species does not guarantee that they will respond identically to some ecological factor such as temperature or light. Nearly every elementary biology textbook has illustrations of geographical variation in the size or color of some animal: there are large specimens from cold regions and small ones from hot regions, dark-colored animals from wet regions and light-colored ones from dry areas. Ecological differences may be more difficult to illustrate than these but they also exist (actually body size and color probably have ecological importance). Two botanists at the University of Missouri conducted an interesting experiment to show ecological variations in different portions of the geographic range of a species. Seeds of a plant called white snakeroot were gathered from widely separated localities and planted in a greenhouse. After the plants had reached a certain size, they were exposed to light equivalent to the long day lengths conducive to flowering in this species. After 120 days the plants grown from North Dakota seeds had produced mature fruits, those from South Dakota seeds had just reached full flower, and those from Kansas had not yet produced flower buds. These differences corresponded very well to the growing season (the length of time between the last frost in the spring and the first in the fall), which is 129 days in North Dakota and 195 in Kansas. Those plants having the responses of the Kansas population would probably be unable to reproduce in the short growing season found in the northern part of the range of the species. Because the plants were all grown under similar conditions we may conclude that the differences seen are genetic, or hereditary. Evolution apparently has produced local populations which are well adapted to do two things in their particular locality: (1) to take advantage of as much time as is available for vegetative growth and food-making; but also (2) to flower and set seed before a killing frost occurs. The different populations of a species which show genetic differences that have ecological significance are known as *ecotypes.*

EXTREME CONDITIONS

The ability of many organisms to acclimate to temperature, the salt content of water, and various other factors is one complication that must be dealt with in going from the laboratory to the field—that is, in explaining the occurrence of organisms in nature on the basis of experimental findings. The existence of ecotypes is another complication. Another fact that must be remembered is that conditions need not be perpetually outside the tolerance limits of a species for the species to be absent.* In other words, you only have to kill a plant or animal once. Six days a week a stream may be well within the tolerance limits of a certain fish but if on Saturday night a factory releases a dose of toxic effluent, the fish will not be found living there. The occasional extreme condition may be more important than the average condition in determining whether or not an organism can exist permanently in a given area.

LIMITING FACTORS AND THE ENVIRONMENTAL COMPLEX

So far we have considered only the organism's response to a single factor of its environment, but the environment of every organism is composed of many factors. The law of limiting factors, stated by a plant physiologist named F. F. Blackman, will often help us to understand a particular case. Liebig's law of the minimum is an earlier, less general statement of the same idea. This law states that when some process depends on several different factors the speed of the process at a given time is limited by the "slowest" factor. The slowest or **limiting factor** may be either too little or too much of something. A process may be limited early in the morning by too little light and limited in the middle of the day by temperatures that are too high.

Good farmers and gardeners understand the law of limiting factors. They know that if their soil is deficient in calcium, adding phosphate will not make their crops grow any better; calcium is the limiting factor and only if calcium is added will growth improve. If enough calcium is added then some other factor will become limiting, perhaps some other mineral

---

*This was pointed out by W. P. Taylor, one of the early students of wildlife management in the western U.S.

such as phosphate. If this is the case then addition of that mineral will now aid growth, even though it had no effect before. The limiting factor may be something other than a mineral; for instance, there may not be enough or there may be too much moisture, or temperatures may be too low or too high.

The law of limiting factors is the basis for the ecologist's concern about phosphates being added to lakes, where phosphate is very often the limiting factor for algal growth. Since the various other materials algae require are generally present in excess (nitrate may sometimes be low) the addition to lakes of phosphates from detergents, lawn or agricultural fertilizers, or sewage, results in increased growth of algae. This has several effects that may be undesirable for some lakes: the increased growth in itself may be a nuisance; once the algae has died and fallen to the bottom, its decay depletes the oxygen supply, and this may kill some of the more desirable fishes; and the undecayed algal remains help build up the bottom of the lake so that it fills in faster. The addition of some other non-limiting material to the lake would be of much less concern, even if that material is used as a nutrient by the algae. Adding calcium to a lake will not usually cause an algal bloom; there is already more calcium there than the algae can use.

The law of limiting factors is simple enough; it is much the same thing as the old saying that a chain is only as strong as its weakest link. However it does not always hold true, because of the *interaction* of different factors. The combined effects of two factors cannot always be predicted from a knowledge of their effects singly. There are cases known in which a surplus of one factor may compensate for a deficiency in another. The upper lethal temperature of the American lobster is 29°C when the oxygen content of the water in which it is living is low, but by increasing the oxygen content, the upper lethal temperature can be raised to 32°C. Two factors may work *synergistically*. In such a situation the combined effect is greater than the sum of the separate effects (a kind of 1 plus 1 equals 5 situation). Two slightly unfavorable factors may, for example, combine to make an area uninhabitable for some organism. One well-documented example of synergism among humans is that of oral contraceptives and smoking. Women over the age of 40 who use the birth control pill have a death rate from heart disease and blood clots slightly higher than among those women who use some other effective method of contraception. But women who use the pill and also smoke have a death rate from heart disease and blood clots much higher than expected.

ENERGY BALANCE

ENERGY AND WORK

The functioning of every organism is based on energy. Don't worry if you don't have a thorough grasp of the concept of energy; almost no one has. You should, however, try to avoid being completely ignorant of the subject, because such ignorance has led to some of our current environmental problems. The conventional definition of **energy** is *the capacity to do work.* *Work* in this sense can be thought of as moving something; the something may be large like a locomotive, but it may also be small like the molecules moved around by an organism to construct a new cell or repair an old one.

In general, work is done when energy is changed from one form to another. If we pick up a ball and put it on a table, *potential energy* in the form of chemical bonds in our body is transformed in a series of steps to the energy of position of the ball (with a considerable loss of energy from our muscles as heat). If the ball rolls off the table and falls back down to the floor, its potential energy is converted to *kinetic energy* and is lost as heat by friction with the air and friction between molecules in the ball and floor.

The basic physical principles describing the transfer of energy are called the first and second laws of thermodynamics. It is sufficient for our purposes to say that the first law of thermodynamics indicates that energy does not vanish. If there is a certain amount of potential energy in a system, and an energy change occurs, then the work done and the heat produced will equal the amount of potential energy lost. The second law of thermodynamics states that no process is 100 per cent efficient in converting potential energy to work; in other words, some of the original energy is always lost as heat.

Because 100 per cent conversion of other kinds of energy to heat is possible, units of heat are convenient measures of energy. Biologists tend to use the *calorie*, which is the amount of heat required to raise one gram of water 1°C, starting at 14.5°C. Because this unit is so small the actual unit used is the Calorie, with a capital *c*, also known as the kilocalorie, which is equal to 1000 small calories. The Calorie, or kilocalorie, is the unit most of us are familiar with from watching our diet. A quarter pound of hamburger, for example, has about 300 Calories. Energy and power may be expressed in other units, such as joules or watts or horsepower, but they all may be converted to kilocalories.

ENERGY IN THE LIVING ORGANISM

The energy that an organism uses for its work comes from the breakdown of organic molecules (or food) within its cells. There are two broad categories of organisms called *autotrophs* and *heterotrophs,* based on how the organisms obtain their food.

Autotrophs, mainly green plants, produce their own foods from simple inorganic materials in the process called *photosynthesis.* It involves the putting together of carbon dioxide and water (plus some other inorganic materials) to form complex organic molecules, especially sugars but also amino and organic acids. Oxygen is also produced. The green pigment chlorophyll and various enzymes must be present in order for this process to occur.

Photosynthesis is usually summarized in chemical symbols as

$$6CO_2 + 6H_2O \xrightarrow{\text{Light energy}} C_6H_{12}O_6 + 6O_2$$

$$\text{carbon dioxide} \quad \text{water} \qquad \text{glucose} \quad \text{oxygen}$$

Such an equation tells us little more than simply stating the same thing in words. It illustrates only the synthesis of one compound, the sugar glucose; summary statements for the other compounds produced in photosynthesis would be different. Also, it tells us nothing about the actual process by which carbon dioxide and water end up as glucose or some other organic compound.

The details of photosynthesis are unnecessary for our purposes. It is enough to say that the process consists of two rather complicated steps, the light reaction and the dark reaction. In the first, radiant energy absorbed by chlorophyll is used to split water into hydrogen and oxygen. The oxygen released in photosynthesis comes from this reaction. In the dark reaction hydrogen is combined with carbon dioxide to produce an organic compound, the exact identity of which varies among different plants. This compound is then treated in different ways to produce many different kinds of organic molecules. During the process part of the energy absorbed by the chlorophyll is converted to the energy of the chemical bonds of the organic molecules produced.

Although we are not concerning ourselves with the physiological details of photosynthesis, they are of importance in ecological research. Two obvious cases come to mind in which detailed physiological information is valuable in understanding some ecological problem. The desert plants referred to as succulents have to some degree uncoupled the light and dark reactions. By so doing they are able to keep their stomata closed in the daytime, when it is hot and water loss would be high, and to open them, taking in carbon dioxide, at night when it is cooler. A second case is the discovery of a new pathway in the dark reaction portion of photosynthesis. Plants using this pathway for making carbohydrates are referred to as C_4 *plants,* while those using the pathway that plant physiologists had previously worked out are known as C_3 *plants.* For several reasons, but particularly because light strongly stimulates respiration (*photorespiration*) in C_3 plants, C_4 plants produce food more efficiently. Ecologically, this means that plants having the C_4 pathway and ecosystems dominated by these plants may be more productive than plants and ecosystems lacking it. Sugarcane, corn, and crabgrass are example of C_4 plants. The C_3 pathway occurs in many plants; these include wheat, sugar beets, rice, and all trees.

Heterotrophs, which include all animals and other organisms such as fungi and many bacteria, cannot make their foods directly from simple inorganic materials. They must take in food that is already formed, by eating plants or other animals that have eaten plants.

The process of **respiration** is basically the same in autotrophs and heterotrophs. Nonbiologists may think that respiration is a fancy word for breathing in the same way that perspiration is a five-dollar word for sweat but this is not the case. Respiration is a process occurring within all living cells whereby organic molecules—carbohydrates, fats, proteins—are broken down to yield energy. In both plants and animals the energy is stored in a phosphorus-containing compound called ATP (adenosine triphosphate). ATP is the immediate source of energy for contracting muscles or doing other kinds of work.

In the most usual type of respiration, *aerobic respiration,* the breakdown occurs through the combination of the organic molecule with oxygen in a complicated series of steps. The connection between breathing and respiration is that breathing is the way the oxygen eventually used in the cells gets into the body. Aerobic respiration may be shown in chemical terms as

$$C_6H_{12}O_6 \; + \; 6O_2 \; \longrightarrow \; 6CO_2 \; + \; 6H_2O$$

$$\text{glucose} \qquad \text{oxygen} \qquad\qquad \text{carbon} \qquad \text{water}$$
$$\text{dioxide}$$

In this reaction energy which goes to produce 38 molecules of ATP is released, as well as an additional amount of energy lost as heat.

There is another kind of respiration known as *anaerobic respiration* or *fermentation.* Chemically, this begins in the same way as aerobic respiration but it does not go as far. The carbohydrate being respired may end up as carbon dioxide and alcohol rather than as carbon dioxide and water. Since there is still energy left in the alcohol, anaerobic respiration is obviously a less efficient process than aerobic respiration. Anaerobic respiration as practiced by yeast may be shown as

$$C_6H_{12}O_6 \; \longrightarrow \; 2C_2H_5OH \; + \; 2CO_2$$

$$\text{glucose} \qquad\qquad \text{ethyl} \qquad \text{carbon}$$
$$\text{alcohol} \qquad \text{dioxide}$$

In this reaction energy which goes to produce two molecules of ATP is released, as well as an additional amount of energy lost as heat. Anaerobic respiration in animal cells produces a compound called lactic acid instead of alcohol. Remember that these chemical statements only show the materials present at the beginning and the end of the reactions; they say nothing about the complex steps that make up the actual process of respiration.

Let us follow the flow of energy through an individual animal (Fig. 2–3). The food that it eats contains a certain amount of stored energy; it may be called the animal's *gross energy intake*. Since the processes of digestion and assimilation are not 100 per cent efficient some of the energy is lost in feces. Also, some of the energy stored in compounds that are digested and absorbed is later given off in various excretions such as urine and sweat. The undigested, unassimilated, and excretory fractions are often lumped together and called *excretory energy*. The energy remaining is essentially all the energy that the animal has to perform all its work. A certain amount of energy is required by the animal just for existence, including such activities as repairing and replacing cells, pumping fluids around in the body, and obtaining, digesting, and absorbing food. Beyond this *existence energy* level, the organism must be able to marshal some additional *productive energy*. The amount by which gross energy intake is greater than existence plus excretory energy is the amount of productive energy available. The productive energy can be used for

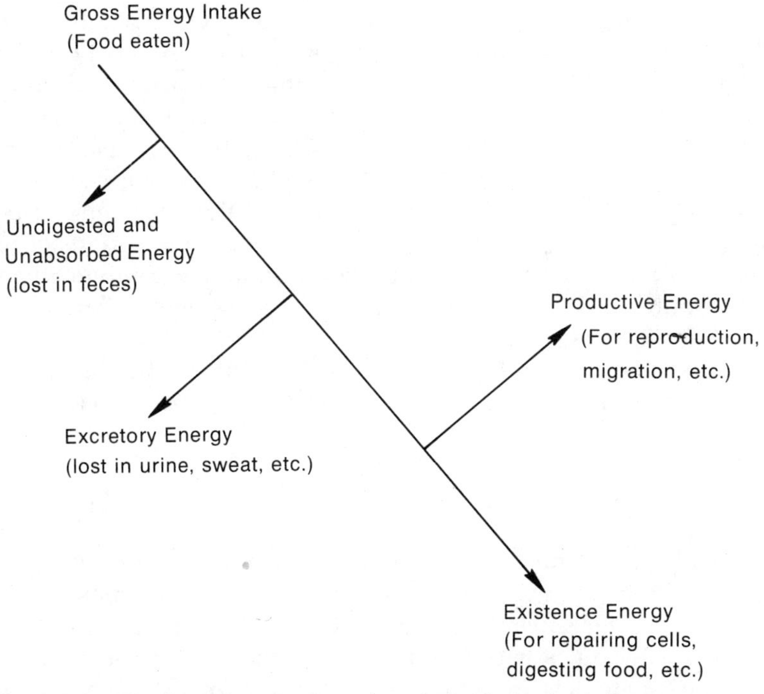

Figure 2–3. The flow of energy through an individual animal. Energy enters as food; some is lost in feces and urine. The rest is used (and eventually given off as heat) or stored (as fat or as other kinds of tissue).

various kinds of work such as growing, mating, nest-building, ditch-digging, or tennis playing. If it is not so used, it accumulates as fat. For birds, this is a normal seasonal occurrence in preparation for migration or for surviving low winter temperatures. For modern man getting fat is the result of the same process of taking in more calories than are used, but the fat rarely serves any useful function.

The importance of a favorable energy balance in the life of an individual organism is clear. A plant that does not fix more energy in photosynthesis than it uses in respiration soon dies. Such a fate may befall a seedling that sprouts beneath a dense forest canopy. A bird that is not able to generate enough productive energy in midwinter will die of starvation or from some malfunction caused by its body temperature falling too far below normal. Sparrow-sized birds do not have enough energy stored as fat to last them more than a day or so without food, depending on how low the temperature is. Circumstances that prevent their feeding for very long are deadly. In the northern part of the range of the bobwhite, storms that cover the food supply with a layer of ice or crusted snow often leave a covey dead on its roost, their stomachs empty.

Energy balance also has implications at the population level. Even if the individual can maintain an energy balance to insure its own survival, the species will not persist in an area unless there is enough additional productive energy available to reproduce—energy to carry on courtship, fight off rivals, produce eggs, and feed young. The University of Illinois ecologist S. C. Kendeigh has pointed out that the various activities requiring heavy expenditures of productive energy tend to be spaced seasonally so that they do not overlap (Fig. 2–4). In migratory birds, for example, the fall molt, in which all the feathers are replaced, is sandwiched between the end of feeding young and the beginning of laying down fat as fuel for the southward flight. There are a few kinds of birds, though, in which the young must be fed until about the time when the birds go south. Included are birds whose feeding methods require the development of considerable skill, such as flycatchers. In many of these species, the molt does not occur in the fall but instead is delayed until the birds complete migration and are on their wintering grounds.

ENERGY SUBSIDIES

Probably all organisms receive **energy subsidies**, that is, energy from outside their own metabolism, to some extent. Gulls use energy from air currents to soar. Organisms in tidal zones use tidal energy to help bring in food and carry out wastes. Lizards and probably even some blackbirds use solar energy to help raise their body temperature. Modern man, however, is uniquely dependent on energy subsidies. One human, depending on size and activity, can get by very well on 2300 kilocalories per day. One color television set has a power demand of 332 watts, or about one

Month	Major Energy-requiring Activities
March	Hunting for scarce food
April	Territory establishment and courtship
May	Nest-building and egg-laying
June	Incubation
July	Feeding young
August	Molting
September October	This seems to be a fairly prosperous time
November December January February	Surviving low winter temperatures

Figure 2–4. Seasonal spacing of the prime energy-requiring activities of a nonmigratory bird. Migratory birds need less energy in the winter for surviving low temperature but have high energy requirements for migration in fall and spring.

third of a kilowatt. Since one kilowatt is equal to 860 kilocalories per hour, your color television set requires close to 290 kilocalories per hour. This means that amount of energy that keeps you alive and functioning 24 hours a day will only keep your television set on for nine hours.

In fact, the average energy consumption per capita in the U.S. is about 230,000 kilocalories per day, or just about 200 times the caloric intake one person needs to live well. Not all of this energy is expended by each person directly, of course. Much of it is used in the manufacture, transportation, and sale of things that are eaten, worn, driven, watched, and discarded. And this figure is not typical of the world as a whole. Nearly everyone has seen the statistic that the U.S., with 6 per cent of the world's population, uses 35 per cent of its energy.

The functioning of the ecosystem, no less than the functioning of individuals and populations, depends on energy. In a later section we will consider energy flow at the ecosystem level.

ANIMAL BEHAVIOR

The scientific study of animal behavior is called *ethology.* It is concerned with what animals do in their daily lives, the functions of their various acts, and the evolution of behavior. Konrad Lorenz, Niko Tinbergen, and Karl von Frisch won the Nobel prize (in physiology and medicine) in 1973 for their pioneering studies of animal behavior.

Much of the ecology of an animal is rooted in its behavior; where it roosts may determine how cold a climate it can live in; its feeding behavior is the basis of predator-prey relationships and a link in a food chain; and habitat selection, territoriality, and many other important ecological topics have behavior as their basis. Much behavior can be considered under the heading of individual ecology, such as maintenance behavior (preening, bathing, sleeping, etc.). However, a very large part of the behavior of vertebrates and some invertebrates is social behavior, and properly belongs under the heading of population ecology. Included is sexual behavior (courtship and mating), *agonistic* behavior (fighting, fleeing, and related acts), and care-giving and care-soliciting behavior (parental care).

One of the great contributions of the field of ethology has been the realization that the world of an animal is not necessarily the same as a human's world. An ant's world is different from ours, and not just in the obvious sense of scale. Its perceptions based on its compound eyes, its other sense organs, and its arthropod nervous system are not the same perceptions we have, even if we lie on the ground and look through grass stems. The German word *umwelt* is sometimes used to refer to the surroundings as perceived by a particular species.

Most animals seem to react in a stereotyped way to very specific features of other organisms or their environment. A European robin will defend its territory not only against another male robin but against a tuft of red feathers hung up in the air. In the three-spined stickleback, a small fish, the male will court a glob of clay of almost any shape as long as it is roundish underneath (like a female ready to lay eggs) and is not red (like male sticklebacks). The features of other organisms or the environment that elicit stereotyped behavior of this sort are called **releasers.** In the last example, if the glob of clay is red, it will release territorial rather than courtship behavior.

Mammals, with their well-developed brains, have behavioral patterns in which the innate becomes less important and the learned becomes more so. (The textbook definition of *learning* is the adaptive modification of behavior by experience.) In humans, of course, learning dominates behavior but, even in humans, behavior develops within constraints set by heredity. One fascinating (and controversial) attempt to abstract some of the universally human patterns of behavior from the jumble of cultural patterns is a book entitled *The Imperial Animal* by Lionel Tiger and Robin Fox (New York, Holt, Rinehart, & Winston, 1971).

PROXIMATE AND ULTIMATE FACTORS

When we ask a biological question with a "why" in it—"Why do birds migrate?" or "Why do varying hares turn white in the winter?"—we are usually asking two separate questions with quite different answers. We are asking, on the one hand, what it is in the environment of the organism that makes it migrate or turn white. That is, what features of the environment serve as cues for the organism to respond physiologically or behaviorally in this way? We are asking, secondly, an evolutionary question: what makes it advantageous or necessary for the organism to migrate or turn white? The answers to the first question are *proximate factors,* and to the second, *ultimate factors.* This terminology was first employed by the British biologist J. R. Baker in writing about a question of the same sort, "Why do birds breed when they do?"

For the varying hare, the most likely ultimate factor in its changing color from summer to winter is camouflage. By matching the snow in the winter and the ground in the summer, individuals are more successful in avoiding predators. It is not, however, the presence or absence of snow nor even high or low temperature that causes the hare to molt its brown fur and grow white; it is day length. If we take a brown hare and put it in the laboratory under a short day length of nine hours it will molt from brown to white, and it will do this whether the laboratory is hot or cold. If we give a white hare day lengths of 14 hours it will begin to shed its white fur and brown will grow in. Day length, or *photoperiod,* is acting as a proximate factor. Under natural conditions, of course, the changes in day length would be associated with the change of seasons and the arrival and departure of snow cover and, consequently, the response would synchronize the animal's color with the color of its background.

HABITAT SELECTION

For some kinds of organisms reaching a favorable environment is largely a matter of chance. A dandelion plant may produce a thousand seeds and those that reach a suitable area—bare, sunny soil—will germinate and grow. Other organisms, mainly animals, show behavior that serves to locate them in a favorable environment, in the process known as *habitat selection.*

When we ask why these organisms occur where they do—why the chipmunk lives in forest or why the bobolink lives in grassland—we are again asking two separate questions. We are first asking what is it about grassland that makes it favorable for the bobolink. What are the ultimate factors causing (or allowing) it to live there? But we are also asking what environmental cues cause the bird to establish territories in such a habitat. In other words, what are the proximate factors causing it to live in grassland?

At the level of ultimate factors we may conclude that in this habitat the organisms find what they need to survive and prosper—such as the right kind of food for themselves and their young, suitable nest sites and materials for nests, cover for escape from enemies, and perches to deliver the song that will attract mates and drive away rivals. Some of these may also serve as proximate factors, but some or all of them may not be observable by the animal at the time it is selecting its habitat. The caterpillars on which many birds feed their young may still be eggs if habitat is selected in the spring or they (those that survived) may already be hidden away as pupae if habitat is selected in the fall. The animal may then have to use some other feature of the environment, one that is associated with the essential feature, as a basis for its choice.

A European bird called the lapwing is adapted in its structure and behavior to living in short grassy vegetation. When it arrives on its breeding ground from the south, however, the grass in all open areas is short. Apparently the factor in the spring that determines which meadows the birds will settle in is not grass height but color. The attractive areas are gray-brown; these are unproductive areas in which the grass will be short and sparse in the middle of the summer. Meadows which are already green are more fertile and later in the summer they will support lush vegetation, too tall and thick for these birds. In the spring the birds avoid these areas even though the grass is not too tall at the time they are selecting breeding sites. In this case, then, color of the field seems to be the proximate factor, the environmental cue to which the birds respond. By responding in this way they will usually nest in areas in which vegetation height and density in summer—ultimate factors, factors important in their nesting success—are suitable.

The Scandinavian ecologists G. Svärdson and O. Hildén have suggested that, for highly motile organisms, habitat selection may be a kind of two-stage process in which areas are first visited on the basis of their general appearance, structure, or landscape, with finer details being examined before the organism actually settles in the area (Fig. 2–5). In the peach aphid, a small insect, dispersal is by winged forms (the life history of aphids is too complicated to describe fully but there is an alternation of wingless, feeding forms with winged forms which accomplish dispersal). The winged forms launch themselves actively into the wind, allow themselves to be carried along, and then actively drop out and land on vegetation. An entomologist counted the numbers landing on two small trees, one a peach and one a euonymus. The euonymus is not a host plant for this insect. The numbers landing on the two trees were essentially the same, with more than 10,000 aphids landing on some days. But, at the end of the dispersal season, there were dense populations of aphids on the peach and none on the euonymus, indicating that the many thousands of individuals which landed later moved on. The first step in habitat selection was evidently visual, a response to the sight of the small trees (or perhaps to some more abstract feature like projecting greenness), but there was a second stage involving a choice between peach and other trees that was, perhaps, based on some chemical factor.

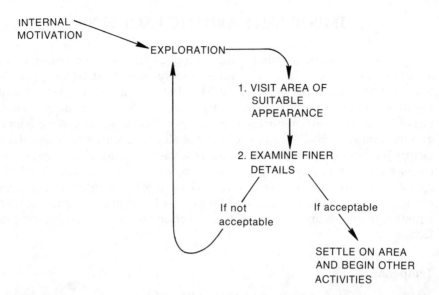

Figure 2–5. Two-stage habitat selection. If finer details of the habitat are acceptable, the animal settles and goes on to other activities, such as establishing a territory, or attracting a mate. If the habitat is unacceptable the animal returns to the first stage of visiting superficially suitable areas discovered in exploration.

One very interesting question is the degree to which the preferred habitat is innately determined and the degree to which it is determined by the early experience of the organism. The answer may be different for different kinds of animals. The very few experiments and observations that have been done with birds and mammals suggest that a strong predisposition for a particular habitat is inherited but that this can be modified somewhat by experience (Table 2–1).

Table 2–1

Two German ornithologists, R. Berndt and W. Winkel, studied banded pied flycatchers to see whether young birds tended to choose the kind of forest they were born in (deciduous forest is considered the typical habitat of the species). They kept track of all the cases of pied flycatchers they had banded as young that were found nesting at least 10 kilometers from where they hatched. What they found is shown below.

Kind of Forest where Hatched	Percentage of Young Breeding in	
	Deciduous Forest	*Coniferous Forest*
Deciduous	69	31
Coniferous	79	21

Pied flycatchers, whether they were hatched and reared in deciduous or coniferous forest, settled if possible in deciduous forest. There seemed to be no learned preference for coniferous forest. Those in coniferous forest probably settled there initially because the deciduous forest areas were already full.

From Berndt, R., and Winkel, W., *J. Ornithol.,* 116:195, 1975.

IMPORTANT ABIOTIC FACTORS

The list of abiotic factors having ecological effects on some organism at some time is very long. The more widely important factors include temperature, moisture, light and other kinds of radiation, texture and chemical composition of soil (for many kinds of plants and soil animals), dissolved gases and chemicals (for many kinds of aquatic organisms), gravity, pressure, and sound. The study of all the specific effects that such factors have upon organisms constitutes a vast subfield of ecology, overlapping the field of physiology, which is referred to as physiological ecology. We will discuss here only a few of the ecological interrelationships of organisms and abiotic factors. (Abiotic factors especially important to aquatic organisms are discussed in the sections on aquatic ecosystems in Chapter 5.)

Temperature

Temperature is important in controlling the rate of processes inside an organism and, thereby, its activity. In general, most important internal reactions proceed more quickly at a higher body temperature. This is true for all organisms, but there is a fundamental difference among organisms in the relationship of body temperature to environmental temperature.

In one group, *poikilotherms*, body temperature tends to match environmental temperature. In the other group, **homoiotherms** or *homeotherms*, body temperature tends to stay constant even when the environmental temperature changes. The situation is actually a little more complicated. Homoiotherms, or warm-blooded animals, are mainly birds and mammals. They have automatic, basically physiological mechanisms for keeping body temperature constant despite changing outside temperatures. These mechanisms differ somewhat from one kind of animal to another but are basically similar. Keeping body temperature high at low environmental temperatures involves increasing metabolic rate (which is the same thing as increasing heat production) and, if possible, increasing insulation. Most homoiotherms accomplish this by fluffing up feathers or fur to increase air space. Keeping body temperatures low at high environmental temperatures involves lowering heat production and increasing heat loss from the body by various means, such as evaporating water from sweat glands or by panting.

Poikilotherms, cold-blooded animals and plants, have no internal physiological means of keeping body temperature constant; if you put a toad in a refrigerator, its body temperature goes down. Even though this is true, if you were able to take the body temperature of a bird and a toad at various times during a day while they were free to move around, you would find that the body temperature of the toad was not very much more variable than that of the robin. The toad, however, uses behavioral rather than physiological means of regulating its body temperature. When the environmental temperature is low the toad hops around more, generating heat from the work done by its muscles, and stays in the sun. When it is hot outside, the toad stays quietly in the shade. Other poikilotherms have different methods of regulating body temperature. If you continue taking the toad's temperature on a cold night, you will find that its temperature drops several degrees and, if you continue as winter sets in, its temperature drops still further until it finds a hole for hibernation and settles in. During this time, you would find that the bird's temperature stays practically constant.

The relationship between the organism and its environment is very different in homoiotherms and poikilotherms (Fig. 2–6). The way in which extremely low environmental temperatures are tolerated, for example, is different. For poikilotherms, which begin to become inactive as temperatures drop, it is largely a matter of avoiding freezing. They may do this by hibernating in the ground below the frost line. Often they will build up chemicals in their body fluids that lower the freezing point (like antifreeze in a car radiator). Plants cannot dig down to where it is warmer and cellular changes, both chemical and physical, are especially important. And they are effective. Ponderosa pine needles in midwinter can survive temperatures well below −60°C with no injury. In experiments, the bark of redosier dogwood can be acclimated to withstand temperatures of −196°C.

For homoiotherms, surviving low temperatures is approximately a matter of getting enough food to keep heat production high enough to

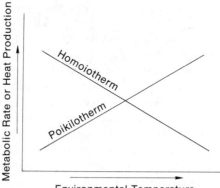

Figure 2–6. The relationship between metabolic rate and environmental temperature in homoiotherms and poikilotherms. Metabolic rate may be measured in units of heat production or in units of carbon dioxide produced or oxygen used. In homoiotherms, over a wide range of temperatures, metabolic rate increases as environmental temperature decreases, with the result that body temperature remains about constant. In poikilotherms decreased environmental temperature leads to decreased body temperature (unless the animal can compensate by behavioral means) with a consequent decrease in metabolic rate.

maintain normal body temperature. If a homoiotherm cannot find enough food (for example, a bobwhite during an ice storm) its body temperature will drop and it will die quickly at a body temperature well above freezing.

It may not be clear at this point what advantages there are to homoiothermy. The homoiotherm is condemned to a life of running around, eating continually to get enough food to metabolize the energy to keep running around. Why isn't the life of a poikilotherm better? On a chilly day, its body temperature stays too low to allow it to be active but at the same time its bodily processes slow down so it doesn't require much food. In the winter, when food is scarce anyway, doesn't it make more sense to be snug (although comatose) with a body temperature of 4 or 5°C and a metabolic rate so slow that a few grams of fat provide enough energy to last the winter? Even if homoiothermy has its good points, why maintain such high temperatures, so near to the upper lethal limit and requiring so much food? Why don't birds, rabbits, humans have a body temperature of 75 or 80°F (24 or 26°C)? The answers to these questions are ultimately ecological and evolutionary, like the answers to all biological questions. The main point is that organisms do not live alone; they live with others that are predators and competitors. On a chilly day while the poikilotherm sits in its burrow, the homoiotherm competing with it has eaten the food that the poikilotherm might have eaten the next day and has, in the meantime, moved on to another source of food. A homoiotherm with a body temperature of 80°F (26°C) might fare well enough browsing sluggishly by itself but, when it is alongside another animal having a body temperature of 98.6°F (37°C), it is the latter that is fast enough to elude a predator and browse again.

There is one group of organisms that has managed to combine some of the advantages of homoiothermy and poikilothermy. These are the basically warm-blooded animals that hibernate in the winter, the *heterotherms*. As winter approaches, the homoiotherms that hibernate find a place to stay and become essentially poikilothermic so that as the temperature in their hibernaculum decreases so does their body temperature. They save the energy they would otherwise spend in keeping their body temperature constant and in moving about. Their **hibernation** is, however, a more controlled state than that of poikilotherms. If the temperature in their hibernaculum gets close to freezing their heat production begins to increase. If the temperature goes still lower they rouse, become thoroughly homoiothermic again and, if they are lucky, find a deeper hole.

Hibernation is so much a part of folklore and common knowledge that we do not always realize that only a few mammals use this means of coping with the low temperatures and diminished food of winter. Woodchucks, 13-lined ground squirrels, and jumping mice hibernate. Raccoons, possums, foxes, and squirrels do not. Not even bears and chipmunks hibernate, although they spend the winter in a burrow. They sleep much of the winter, but their body temperature does not drop much more than ours does when we sleep. The bear is a large animal with a slow metabolic rate and can store enough fat to sleep through the winter without becoming poikilothermic. The chipmunk stores food to eat when it wakes up every so often. Very small animals like shrews and most mice do not hibernate; they would have a hard time accumulating enough fat to get through the winter as hibernators and, anyway, it is not very wintery where they live under the insulation of leaves and snow.

From ancient times until the eighteenth century many people believed that hibernation was widespread in birds. This idea fell into disrepute as the scientific study of birds increased and yielded only observations of migration and none of hibernation. It was not until the winter of 1946, when the desert biologist E. C. Jaeger found a poor-will hibernating in a rock cranny in the Chuckawalla Mountains of California, that the idea of hibernation in birds was restored to scientific respectability. It is clear, nevertheless, that hibernation as a method of coping with winter has been favored by natural selection even less in birds than in mammals.

The **"life form"** system of categorizing plants was devised by the Danish ecologist C. Raunkiaer and is based primarily on the methods by which plants survive the coldest season. Specifically, the main categories are based on the location of the bud from which new growth will sprout. As modified by the Swiss plant sociologist J. Braun-Blanquet and by others, the system now generally includes the broad divisions shown in Table 2–2.

The proportions of the flora in the various categories (which Raunkiaer called the "biological spectrum") varies from one climate to another: deserts typically have a high proportion of therophytes; prairies and steppes have a preponderance of hemicryptophytes; alpine areas may have a high proportion of chamaephytes (although hemicryptophytes may be even more important) which are well-protected through the winter by deep snow. Tropical rain forest is, of course, typical of a phanerophytic climate. Other factors besides regional climate affect the biological spectrum; in temperate deciduous forests, among the herbs, hemicryptophytes predominate in the dry oak forests and geophytes predominate in the mesic beech-maple forests.

Moisture

Whether a piece of land is wet or dry for the organisms living there depends mainly on two factors, soil moisture and the evaporative power of the air. The evaporative power of the air depends on humidity, temperature, and wind. The evaporative power of the air is increased and an area made drier, as far as organisms are concerned, by lower humidities, higher temperatures, and more wind.

Table 2–2 RAUNKIAER'S "LIFE FORM" SYSTEM*

Life Form	Method of Surviving Unfavorable Season	Examples
Therophytes	Annual plants which survive the winter (or dry season) as seeds. Seeds are typically much more resistant to cold and drought than are growing plants.	Annual ragweed, blue-eyed mary, lettuce
Hydrophytes	Rooted water plants. The bud that produces the next year's growth is under water during the winter and is thereby insulated.	Pondweed, water lily, pickerelweed
Geophytes	Next year's bud buried in the soil and thus well insulated. Includes plants with underground stems, such as bulb plants.	Tulips, ferns, trout lily
Hemicryptophytes	Buds are located close to the surface of the ground; accordingly, they may be insulated by litter or snow but do not have the further insulation of soil that geophytes have.	Goldenrods; perennial grasses; all biennials such as carrot (wild and tame); mullein
Chamaephytes	Here the renewal buds are located above the ground but not very high (no more than 25–30 cm). The buds may sometimes be insulated by a snow cover and they are low enough so that they are not exposed to strong winds.	Trailing and creeping shrubs such as wintergreen; fruticose lichens; succulents
Phanerophytes	Mainly trees and shrubs; here the buds are located on shoots more than 25–30 cm above the ground and are thus more exposed than any other group.	Essentially all trees and shrubs; also lianas (climbing vines) such as poison ivy; in the tropics are some herbaceous phanerophytes

*In addition to these major divisions, the classification also includes phytoplankton (e.g., aquatic algae), phytoedaphon (e.g., soil algae), endophytes (mainly plants which are internal parasites), and tree epiphytes (e.g., Spanish moss and many tropical orchids).

Humidity is defined as moisture in the air in the form of vapor; it may be expressed in several ways. It can be given as the actual mass of water in a certain volume of air, *absolute humidity*. A more familiar measure is *relative humidity*, which is the actual absolute humidity expressed as a percentage of what the absolute humidity would be if the air were saturated with water vapor at the same temperature and pressure. The amount of water vapor that air can hold increases with temperature. At −30°F (−34°C) one cubic foot of air can hold only about 0.1 gm; at 40°F (4°C) it can hold nearly 3 gm; and at 100°F (38°C) it can hold nearly 20 gm. If the amount of water stays constant, lowering the temperature raises the relative humidity and raising the temperature lowers it. This is why the air in your house in the winter is very dry (unless you have a humidifier); the cold air outside may have a relative humidity of 80% but actually very little water. Bringing this air into the house and heating it to 68°F (20°C) may drop the humidity to a desert-like 10 or 20%.

Plants and animals must maintain a favorable water balance; that is, they must take in about as much water as they lose and use. For terrestrial plants, water is absorbed from soil water through the roots. The plant uses water in photosynthesis, the hydrogen ending up in carbohydrates, and it loses water to the air through its leaves (*transpiration*). Most of the loss is not through the leaf surface but through microscopic pores called *stomata* (Fig. 2–7). In most plants these are open during the daytime and closed at night; they are also closed when water balance becomes unfavorable and the plant wilts. Less than 1 per cent of the water taken up by the roots is used in photosynthesis; the rest is transpired. Presumably the function of the stomata is to allow carbon dioxide and oxygen from the air to enter the leaf. The water loss is a by-product of this arrangement necessary for photosynthesis and respiration. Consequently, botanists have traditionally regarded transpiration as a necessary evil (or, as the University of Illinois plant ecologist A. G. Vestal is supposed to have said, a damned shame). There is evidence, however, that the cooling effect of transpiration is useful to plants in some circumstances.

Xerophytes, mesophytes, and **hydrophytes** are categories of plants based on their water relationships. Xerophytes grow where it is dry (*xeric*), hydrophytes grow where it is wet (*hydric*), and mesophytes grow where there is a moderate amount of water (*mesic*, meaning medium). For xerophytes (and, during drought, for mesophytes) maintaining water balance poses a problem. For hydrophytes the problem is not too much water but too little carbon dioxide or oxygen, since the concentration of these

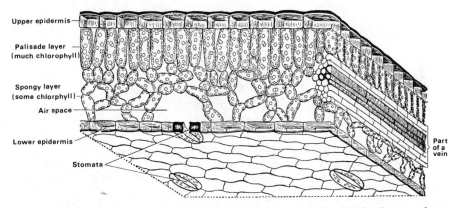

Figure 2–7. A diagrammatic view of a small part of a typical leaf blade of a mesophyte. The front consists of a cross section through the leaf. The side is a longitudinal section. The bottom is a view of the lower surface of the leaf. Water loss from the leaf is mainly through the small pores (stomata) in the lower surface. They are each surrounded by two cells (guard cells) that close the stomata at certain times (usually at night and when the plant is wilted). The stomata open into air spaces in the spongy layer. Water is transported up from the roots through the stems and into the leaves in conducting tissue like that shown in the vein at the right. (From Edmund W. Sinnott, *Botany Principles and Problems*, New York, McGraw-Hill, 1946, p. 110.)

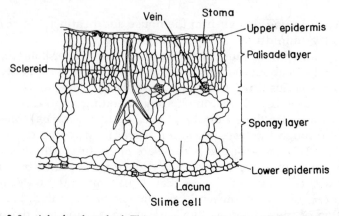

Figure 2–8. A hydrophyte leaf. This is a cross section through a small part of a water lily leaf, which floats on the water surface. The stomata are in the upper surface, the only part in contact with air. The large spaces in the spongy layer, called *lacunae*, do not open through stomata to the outside. They provide buoyancy; they also allow carbon dioxide to accumulate at night to be used by day, and oxygen to accumulate in the daytime for night-time use. The upside-down Y is a thick-walled structure called a *sclereid*. It gives support to the leaf without giving it side-to-side rigidity, allowing it to ripple with wave movements. There is little water-conducting tissue. (From J. E. Weaver and F. E. Clements, *Plant Ecology*, New York, McGraw-Hill, 1938, p. 431.)

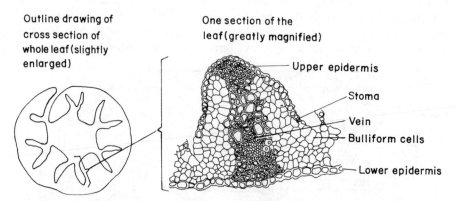

Figure 2–9. Many xerophytic grasses have leaf structures similar to that of western wheat grass, shown here. The stomata are located on the top of the leaf in deep channels. Presumably the air in these channels is less apt to be swept away by wind and, accordingly, is more likely to maintain a high humidity and consequently lower transpiration water loss. The lower epidermis is covered with thick-walled cells through which little water can escape. Under moist conditions the leaf is flat like any grass blade in a lawn. With drought, however, thin-walled cells *(bulliform cells)* on the upper surface of the leaf lose water rapidly, become limp, and allow the leaf to roll up. When this happens the stomata are totally enclosed within the inrolled leaf and the only surface of the leaf that is exposed is the heavily cutinized lower epidermis. The thick-walled cell surrounding the vein are like I-beams, preventing the leaves from drooping and bending the water-conducting tubes. Notice also that the amount of water-conducting tissue is large compared with that in the mesophytic leaf, Figure 2–7. (From J. E. Weaver and F. E. Clements, *Plant Ecology*, New York, McGraw-Hill, 1938, p. 450.)

vital gases is lower in water than in air. A water lily leaf has big open spaces (*lacunae*) inside it where the carbon dioxide given off when the cells respire and the oxygen given off in photosynthesis can accumulate and be recycled (Fig. 2–8).

There are several kinds of xerophytes, each adapted in its own way to life in a dry environment (Fig. 2–9). To botanists, the *true xerophytes* are perennials, usually shrubs such as sagebrush, that have small, hard leaves which can decrease water loss greatly when the soil gets dry and which are dropped altogether during prolonged drought.

Succulents are another kind of xerophyte. Cactuses are examples. They have fleshy tissue in which water, when it is available, can be stored for later use. *Desert ephemerals* are annuals that look much the same as their relatives living in moister regions. Their adaptation to desert life is in their life history rather than in their structure. When enough rain falls, the seeds germinate, the plant grows up, flowers, sets seed, and dies, all in a short period when conditions are fairly moist. The new seeds then lie in the ground until enough rain falls again; this may occur the next year or not for several years.

Another category of xerophytes is the *phreatophytes* (from a Greek word meaning well). These plants grow in the deserts but have roots that go down far enough to tap the water table. Mesquite is a phreatophyte which may have a root system reaching down 175 feet.

Water conservation is serious matter for xerophytes; it is not, however, their central problem. That, as for every organism, is success in a community of competitors and predators. The Russian plant physiologist N. A. Maximov noted that xerophytes tend to have larger bundles of water-conducting tissue in their stems and leaves than do mesophytes. Clearly, this is no adaptation for conserving water; in fact, it allows faster water loss. Consider a xerophyte and a mesophyte growing near one another in the desert and drawing on similar amounts of soil moisture. In the heat of the day the mesophyte wilts because it cannot transport water as fast as it is lost. The wilted plant can no longer carry on photosynthesis, and eventually it dies without having thoroughly depleted the soil moisture around it. The xerophyte remains unwilted and carries on photosynthesis until the moisture around its roots is gone; only then are the water-conserving mechanisms used. In the meantime it has stored food and possibly produced flowers and seeds.

We can imagine the perfect water conserver: possibly, it would be ball-shaped, with a thick waxy covering. It might survive for a long time—until perhaps a mule deer comes along and eats it. But meanwhile some less conservative plant will have reproduced and populated the desert.

For animals the sources of water gain and loss are somewhat more complicated. The usual sources of gain are drinking water, water in food, and metabolic water (water molecules produced in the breakdown of foods). Loss is through urine, feces, and the water evaporated from the skin and lungs. There are two kinds of adaptations in animals living in arid regions. One is lowering water loss as much as possible. This is achieved by reabsorbing water in the large intestine to produce feces as nearly dry as possible and by using only the minimum water necessary to excrete the nitrogenous wastes of the urine. Cutting water loss from the lungs and skin is a dilemma for desert animals similar to reducing transpiration loss for plants. Water loss from the lungs is an inevitable consequence of breathing; this water loss, along with that from the skin, also keeps the animal from becoming too hot. About the best that can be done is for animals to become nocturnal and, if they are small enough, to stay in burrows during the day. Inside the burrow it is cooler and the humidity becomes high enough that there is less evaporation from the lungs.

The second means of adapting to life in arid regions is the ability, when the animal does suffer a temporarily unfavorable water balance, to tolerate the resulting dehydration and elevated body temperature. Camels can tolerate water loss exceeding 25% of their body weight. Man can tolerate a loss up to about 20% but, by the time he has lost about 10%, he is physically and mentally unable to take care of himself.

For aquatic animals the corresponding problem is one of maintaining *osmotic concentration*, keeping a proper salt balance. For marine invertebrates there is rarely a problem since most have a salt concentration similar to that of seawater (they are said to be *isotonic* with seawater). Freshwater organisms, whether invertebrates, fish, or amphibians, are *hypertonic* to the water in which they live (they have a higher salt concentration) and, consequently, must have ways of actively absorbing salts and also of recovering salts from the urine before it leaves the body. Marine vertebrates are *hypotonic* (a lower salt concentration) to seawater and thus must be able to excrete a concentrated urine to get rid of excess salt that is taken in with the seawater or in the food they eat. Seawater contains 3% salt. Whales can excrete urine more concentrated than this, but the best humans can do is 2.2%. Marine birds and reptiles have an additional organ that gets rid of excess salt; this is a pair of nasal glands that lie on the top of the head just above the eyes and open into the nasal cavity. These glands secrete sodium chloride (table salt) in concentrations up to about 5%.

The interaction of moisture and temperature is sometimes important in the life of an organism. A graphical method of studying such interaction, the *climograph,* is shown in Figure 2–10.

Light

Light is a kind of radiation, specifically the visible wavelengths of electromagnetic radiation. This portion of the spectrum runs from violet (short wavelength) to red (long wavelength) (Fig. 2–11). Wavelengths just shorter than violet are ultraviolet radiation, which is visible to insects but not to us, and X-rays. Wavelengths just longer than red are infrared and

Figure 2–10. One method of considering temperature and rainfall at the same time is in a graph such as this one, called a *climograph*. It may be used in a purely descriptive way just to see the difference, for example, between the climate of rain forest and desert. It may also be used as a tool for prediction. In this climograph the climate of the best European range of the Hungarian partridge is compared with that of Havre, Montana, and Columbia, Missouri. The partridge was introduced successfully in Montana but failed in Missouri, implying that hot (or hot and wet) summers are limiting out that low winter temperatures are not. Future introductions (if justified) should not be in areas having a climatic regime such as that found in Columbia, Missouri. (From E. P. Odum, *Fundamentals of Ecology*, Philadelphia, W. B. Saunders, 1971, p. 125.)

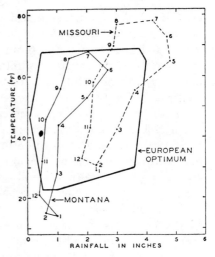

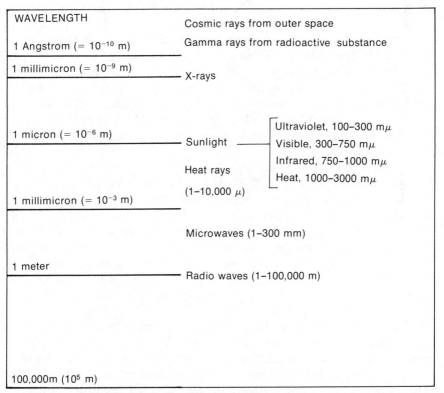

Figure 2–11. Electromagnetic radiation ranges from the very short, high-energy cosmic rays to the very long, low-energy radio waves. Visible light occupies a small part of the electromagnetic spectrum with wavelengths between about 0.30 and 0.75 μ (microns).

microwave radiation. Ecologically important light is generally sunlight, or solar radiation. Solar radiation contains wavelengths other than visible light, of course. Somewhat more than half the energy in solar radiation reaching the earth's surface is in the visible range and somewhat less than half is in the near infrared (the infrared wavelengths near the visible spectrum). Because of the ozone layer of the atmosphere, only a little ultraviolet radiation reaches the earth's surface.

Light, or solar radiation in general, has many effects of ecological importance ranging from the role of light (mainly red and blue) in photosynthesis to the role of ultraviolet radiation in vitamin D production in animals. At this point we will deal with only two ecological aspects of light, shade tolerance and photoperiodism.

Shade tolerance is the ability of a plant to survive and grow in the shade. We think of shade as meaning lowered light intensity, which it does. The canopy of a beech-maple forest may screen out over 99% of the sunlight hitting it. In the middle of a summer day, the light intensity outside the forest may be over 10,000 footcandles and inside the forest less than 100 footcandles. Shade also means other things, however, including reduced wind, high humidity, and moderated temperature. Still, the ability to stay alive and grow in low light levels is probably the one most important feature of shade tolerance.

Foresters classify trees in five categories of shade tolerance, from very intolerant to very tolerant (Table 2–3). This knowledge is useful in managing forest lands and also simply in gardening. A shaded garden is of very little use unless one plans to grow native wildflowers or their counterparts from elsewhere in the world. Plants from which you obtain an edible crop are almost all shade intolerant, perhaps because they need high levels of sunlight to make enough food for themselves plus enough for man to harvest.

Photoperiodism refers to the length of the light and dark portions of the 24-hour day. Because of the way in which the earth's axis is tilted, the length of day and night changes seasonally everywhere except at the equator (Fig. 2–12). The longest day occurs about June 21 and is called the summer solstice; the shortest day is December 21, the winter solstice. The spring and autumn days when day and night are each 12 hours occur on March 20 and September 23, the vernal and autumnal equinoxes. Where I live, 42° north of the equator, the longest day is about 15 hours long and the shortest about 9. Further south the difference diminishes and further north it increases.

For most of the earth, the change in day length is the most accurate information available to an organism on the advance of the seasons. For many organisms, day length, or photoperiod, is an important factor in the scheduling of their life history events. Most birds of the temperate regions begin to come into breeding condition not because of increased temperature or a better food supply but because of a lengthening photo-

Table 2–3 SHADE TOLERANCE OF AMERICAN FOREST TREES*

Eastern Conifers		Western Conifers	
Very tolerant		*Very tolerant*	
Eastern hemlock	Balsam fir	Western hemlock	Pacific yew
Tolerant		Alpine fir	California torreya
Red spruce	White spruce	Western redcedar	
Black spruce	• Northern white-cedar	*Tolerant*	
Intermediate		Sitka spruce	Redwood
Eastern white pine	Baldcypress	Engelmann spruce	Incensecedar
Intolerant		Mountain hemlock	Port Orford white-cedar
Eastern redcedar	Shortleaf pine	Pacific silver fir	
Red pine	Loblolly pine	Grand fir	Alaska yellow-cedar
Pitch pine	Virginia pine	White fir	
Very intolerant		*Intermediate*	
Tamarack	Longleaf pine	Western white pine	Douglas fir
Jack pine		Sugar pine	Bigcone-spruce
		Monterey pine	Red fir
Eastern Hardwoods		Blue spruce	Giant sequoia
Very tolerant		*Intolerant*	
Eastern hophorn-beam	American holly	Limber pine	Knobcone pine
American horn-beam	Sugar maple	Pinyon	Bishop pine
American beech	Flowering dogwood	Ponderosa pine	Bigcone spruce
		Jeffrey pine	Noble fir
Tolerant		Lodgepole pine	Junipers
Red maple	Tupelos	Coulter pine	
Silver maple	Persimmon	*Very intolerant*	
Box elder	Buckeyes	Whitebark pine	Digger pine
Basswood		Foxtail pine	Western larch
		Bristlecone pine	Alpine larch
Intermediate			
Yellow birch	Rock elm	**Western Hardwoods**	
American chestnut	Hackberry	*Very tolerant*	
White oak	White ash	Vine maple	
Red oak	Green ash	*Tolerant*	
Black oak	Black ash	Tanoak	Madrone
American elm		Canyon live oak	California laurel
Intolerant		Bigleaf maple	
Black walnut	Sweetgum	*Intermediate*	
Butternut	Sycamore	Golden chinquapin	California white oak
Pecan	Black cherry	Oregon ash	Oregon white oak
Hickories	Honey locust	*Intolerant*	
Paper birch	Kentucky coffeetree	Red alder	
Yellow-poplar	Catalpas	*Very intolerant*	
Sassafras		Quaking aspen	Cottonwoods
Very intolerant			
Most willows	Cottonwoods		
Quaking aspen	Black locust		
Bigtooth aspen	Osage orange		

*Adapted from Baker, F. S., A revised tolerance table. *J. Forestry*, 47:180–181, 1949. Some trees are difficult to place in one category with assurance (some seem to behave differently in some regions of the country, for example).

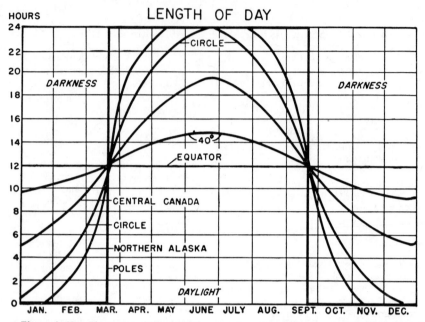

Figure 2–12. Photoperiod through the year at various latitudes. At 40° north, approximately the latitude of Indianapolis and Denver, day length is slightly below 10 hours on January 1 and slightly less than 15 hours at the summer solstice toward the end of June (From A. M. Woodbury, *Principles of General Ecology*, New York, Blakiston, 1954, p. 122.)

period. At some point, temperature or food supply or some other factors may become important in determining just when egg-laying actually begins.

Among plants the timing of many events, including flowering, may be based on photoperiod. *Short-day plants* require days shorter than some maximum length to come into flower; if days stay longer than this the plants will continue growing vegetatively and will not produce flowers. Examples of short-day plants are lespedeza, chrysanthemum, soybean, poinsettia, and violet. The critical day length varies from one kind of plant to another; usually it is somewhere between 11 and 14 hours. *Long-day plants* are plants which need day lengths longer than some critical minimum to come into flower. Here again, the shortest day that will stimulate a long-day plant to prepare for flowering varies between species but is usually 10 to 13 hours. Some long-day plants are red clover, evening primrose, dill, spinach, timothy, and wheat grass. There are also day-neutral plants that will flower under a variety of photoperiods if other conditions are favorable, such as buckwheat, cucumber, impatiens, and corn.

If you think at this point that long-day plants flower on photoperiods longer than 12 hours and short-day plants flower on photoperiods less than that, then you have not read the definitions carefully enough. It should be clear that, depending upon the species, both long- and short-day plants may be induced to flower by photoperiods of 12 hours, or 11 hours, or 13 hours (among others). This points up the fact that photoperiod, in itself, is rarely important to plants and animals as an ultimate factor. Rather, photoperiod is used as a timing device (that is, as a proximate factor) to schedule an activity at the appropriate season.

Most biologists live in a fairly small part of the globe, the forested and grassland parts of the temperate regions, where photoperiod change is pronounced and the main seasonal change is winter to summer. In other parts of the earth, such as the tropics and some deserts, photoperiod may change little or a change from dry to wet season may be the important seasonal change. For these regions natural selection has favored factors other than (or in addition to) photoperiod as the environmental cue for when to breed or flower (Fig. 2–13). In some desert birds courtship and mating may begin within hours after a substantial rain. Here, something about the rain itself seems to be the trigger. In other birds, there is evidence that chemicals that inhibit reproduction become concentrated in the plants eaten by these birds during the dry season. With the coming of the rains, rapid plant growth evidently lowers the concentrations of these chemicals, reproduction is no longer inhibited, and the birds begin to breed.

The first biologists working with photoperiod thought that it was the length of the light period that was important in producing a response. Later experiments showed that long-day organisms kept on a short-day photoperiod would respond as though they were really getting long days if the nights were interrupted even with just a few minutes of light. From this, it was concluded that night length rather than day length was important. Further research, using photoperiods that could never occur in nature, such as 6 hours of light and 36 hours of darkness, have shown that the explanation is more complicated. Very briefly, it appears that organisms have a daily rhythm (a *circadian* rhythm, meaning one of about 24 hours) of susceptibility to light stimulation. During part of the 24-hour day they get no stimulatory effects from light; during the rest, light is

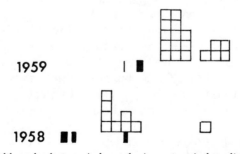

1959

1958

Figure 2–13. Although photoperiod may be important in breeding of the Abert's towhee, the immediate cue for nesting seems to be a factor associated with heavy spring rains. Nesting begins ten days to two weeks after a heavy rain, as early as February or as late as April. The diagram indicates nesting for a mesquite woodland area near Tucson, Arizona. The black bars represent rains of 2.5 cm or more; the boxes represent new nests started at five-day intervals. In 1959 nesting was late, following the only period of heavy rain of the spring; in 1958 nesting was early, following heavy rains in early March. (From J. T. Marshall, Jr., *Proceedings of the 13th International Ornithological Congress*, Vol. No. 2:620, 1963.)

stimulatory. It is the light falling in this phase of the cycle that produces the effect, whether the light is a continuation of a long day or merely some brief period during the night.

Soil

Soil is the loose surface material of land in which plants grow. It cannot be placed in the same category as temperature or light as a simple "factor" affecting organisms, since soil is itself a complex system. It is composed of fragments of the parent mineral material, organic matter in various stages of breakdown, soil water and the minerals and organic compounds dissolved in it, soil gases in the spaces not containing water, and living organisms.

Soil plays obvious roles in plant growth. From it comes the water plants use in transpiration and photosynthesis, the calcium, nitrate, phosphate, and other mineral nutrients that the plant requires in the manufacture of various organic compounds, and the oxygen that the cells of the roots need for respiration. But the soil is a great deal more than this. When a plant or any organism dies the process of decay occurs in the soil through the activities of soil organisms, especially bacteria and fungi. *Humus,* finely ground organic matter mixed with the mineral part of the soil, is produced, and eventually the minerals originally taken in by the plant roots are returned to the soil. The soil is home to many types of animals, from moles to salamanders to earthworms to beetles to protozoans. The number of individuals of some of the smaller invertebrates is incredible to anyone who has never used the special techniques for studying soil organisms. In a patch of soil one meter on each side, or about one yard square, there may be 10,000 small insects called springtails, 100,000 small relatives of the spider called mites, and 1,000,000 thin white roundworms, or nematodes.

The formation of soil, whether from solid rock or from mineral material deposited by a glacier, wind, or water, is a complex process. *Mechanical weathering,* such as the cracking of big rocks into little ones by the wedging action of water freezing in a crack, is important. *Chemical weathering* is also important, especially the leaching downward of alkaline materials such as calcium salts by rainwater containing dissolved carbon dioxide. Also important are the activities of organisms, such as the aggregation of soil particles produced by their passage through an earthworm's gut.

Because soil-forming processes tend to act from the top down, soil develops a vertical structure referred to as the *soil profile* (Fig. 2–14). Soil scientists recognize three main layers, or *horizons.* The uppermost layer is topsoil, or the A horizon, where most of the plant roots are located. Dead organic matter is added to the top as *litter* and, as it is partially broken up, mixed as humus with the mineral soil below. Organic material is largely absent from the B horizon and the mineral parent material is

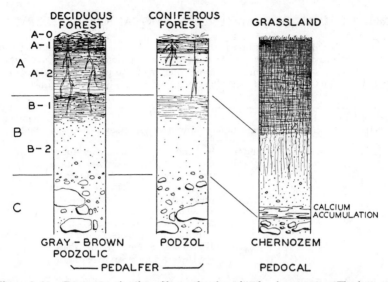

Figure 2–14. Diagrams of soil profiles under three kinds of vegetation. The letters and numbers on the left refer to specific horizons; A-O, for example, refers to dead plant material lying on the surface. The horizons are usually less distinct in grassland soils than in forest soils. "Pedocals" are soils (like grassland and desert soils) from which the calcium (cal) has not been leached. "Pedalfers" are soils (like forest soils) from which much of the calcium has been leached, leaving, along with silicon and oxygen, aluminum (al) and iron (fer). (From Odum, E. P., *Fundamentals of Ecology*. W. B. Saunders Co., Philadelphia, 1971.)

less thoroughly weathered. Materials leached from the A horizon, such as calcium carbonate, may be deposited here. Below this is the C horizon, consisting of more or less unaltered parent material such as glacial drift or bedrock.

What the soil of a particular area is like, the end product of soil formation, depends on many things. Three of the most important are the nature of the parent material, climate, and vegetation (or, better yet, the whole community including animals and microorganisms). For example, where the parent material is a rock called serpentine, the soil which develops is deficient in calcium and usually high in magnesium and nickel. On such areas in the southeastern U.S., Sweden, and elsewhere grow a collection of species very different from those on adjacent areas formed from different types of rock.

Rainfall, evaporation, and temperature strongly influence soil development, so that the arctic, tropic, temperate, and arid regions all tend to develop different types of soil (Table 2–4). In the northern Midwest, however, coniferous forest, deciduous forest, and grassland grow within a few miles of one another and the soils, all derived from the same basic parent material, may be quite different owing to the differing effects of the vegetation. Under the coniferous forest may be a strongly acid soil with a heavy layer of undecomposed litter and a hardpan of leached clay in the B horizon. Under the deciduous forest the litter is thinner and

Table 2–4 TYPES OF CLIMATIC-VEGETATIONAL SOIL DEVELOPMENT

The processes described in this table give rise to *zonal* soils which occupy much of the area of a region having a certain climate and vegetation. Various *intrazonal* and *azonal* soils occur in all regions—for example, serpentine soils, hydromorphic soils (such as in bogs), halomorphic soils (salty soils of arid, poorly drained areas), lithosols (bedrock outcrops), etc.

Type of Soil Development	Climate	Vegetation	Important Processes	Kinds of Soils
Gleization	Cold, little rainfall	Tundra	In summer soils remain wet due to poor drainage caused by permanently frozen lower layers (*permafrost*). In winter freezing at top compresses middle layers (*glei*) which are sticky and blue-gray because of iron in the reduced state. Glei is forced up through cracks and thoroughly mixed. Freezing and thawing sorts out rocks on the surface.	Tundra
Podzolization	Cool, fairly moist	Coniferous or deciduous forest	Seen in extreme form only under certain types of coniferous forest. Percolation of acid water downward leaches out carbonates, A horizon becoming acid. If extremely acid, clays may be leached and deposited in the B horizon and form a hardpan. Decomposition is slow under acid conditions and there is often a sharp break between the humus layer and the upper mineral layer, which may be sandy and ashy gray.	Podzols (coniferous forests) Gray-brown podzols (deciduous forests; podzolization less extreme so that soils are less acid and more fertile)
Laterization	Warm, moist	Tropical forest	Clay minerals decompose rapidly and release bases, keeping soil from becoming highly acidic. Humus decomposes rapidly and does not ac-	Lateritic soils

grades into the mineral soil below; the soil is less acid because basswoods, maples, and dogwoods absorb calcium and other minerals from the subsoil, incorporate them into their leaves and stems, and return them to the soil surface each autumn. Under the grassland the soil may be still less acid, possibly almost neutral, and still more fertile because the nutrient pumping action is even more pronounced. The grass roots penetrate deeply, the plants use large amounts of calcium and other nutrients, and each year not only leaves but the entire aboveground part of the plant dies back, allowing decomposition to free the nutrients at or near the soil surface. The grassland soil is dark through the addition of this organic material plus that added to the soil at various depths by the death each year of many fine roots.

Soil texture is based on the sizes of mineral particles making up the

Table 2–4 TYPES OF CLIMATIC-VEGETATIONAL SOIL DEVELOPMENT
(*Continued*)

Type of Soil Development	Climate	Vegetation	Important Processes	Kinds of Soils
			cumulate. Silica fraction is leached and oxides of iron, aluminum, and manganese remain, giving solid reddish or yellowish color.	
Calcification	Warm to cool, not moist	Grass-land, savanna, desert	Leaching not sufficient to carry away calcium carbonates which may accumulate as a hardpan; soil remains neutral to alkaline and clay is not leached. Death of roots and tops adds organic matter and nutrients. Darkness is related to amount of organic matter and thus to rainfall. In desert wind erosion may remove finer particles.	Prairie soils (eastern grasslands where rainfall is sufficient to leach calcium carbonate) Chernozem soils (humid grassland) Chestnut and brown soils (mixed and short grass prairie) Gray desert soils (sagebrush) Red desert soils (shrubs and cacti)

soil, classified as *gravel, sand, silt,* and *clay*. Soils made up mainly of small particles are called *heavy* soils, in which water soaks in slowly but is retained well; such soils have the potential of being very fertile. *Light* soils, such as sandy soils, are well aerated and allow free movement of roots and water but are relatively infertile.

The difference in potential fertility between light and heavy soils results from the way minerals are retained. Such minerals as calcium and magnesium (specifically, most elements that form positive ions, or *cations,* when dissolved) are stored on the surface of particles. Plant roots remove these mineral cations and replace them with hydrogen ions. The potential fertility of a soil depends strongly on *cation exchange capacity,* a measure of the number of sites per unit of soil (usually 100 gm) on which hydrogen can be exchanged for a mineral cation. Clay particles are small and thus provide more surface area in a given weight or volume as compared with the same amount of sand, pebbles, or boulders. Clay soil may have a cation exchange capacity from twice to twenty or more times that of sand.

A soil with a high cation exchange capacity and, consequently, high potential fertility may be actually infertile if most of the sites on the particles are filled with hydrogen ions. Such a soil will be acid and will have little calcium to supply to the growing plant. *Percentage base saturation,* the percentage of the exchange capacity satisfied by calcium, magnesium, and similar elements, measures this aspect of fertility. If percentage base saturation is 50%, then half the sites are filled by basic ions and half by hydrogen ions.

The soil structure that is generally best for plant growth contains both large and small particles, combining the desirable features of good water retention, drainage and aeration, easy root penetration, and potentially high fertility. A soil which has both large and small particles well represented is called *loam,* and loams are almost always productive. Even soils made up mainly of sand or clay can be given a favorable structure by adding organic matter. Knowledge of the importance of humus in soils is not new in ecology. Long ago the pioneer American ecologists John Weaver and Frederick E. Clements wrote: ". . . its physical effects are so marked that when the organic matter present in a soil is very high the distinctions between sands, loams, and clays are practially obliterated." In clays, the particles may be bound together, improving aeration and water movement, by organic compounds released in the breakdown of humus (and also by gums produced by bacteria and blue-green algae and by the activities of fungi and earthworms). Humus has an even higher capacity for holding water and nutrients than does clay; consequently, deficiencies of sandy soil can be largely remedied by the addition of organic matter.

Fire

Many early ecologists in this country regarded fire as a destructive, unnatural phenomenon, associated with man. Until very recently the U.S. Forest Service shared and spread the same view; Smokey the Bear still represents the Forest Service to most of us. Fire certainly is destructive but most ecologists and many foresters now believe that it is a natural part of the environment and that it is a useful, even necessary, tool in managing some ecosystems.

There are basically three types of fires. *Surface* fires sweep rapidly over the ground, consuming litter and herbs, often killing the aboveground stems of shrubs and scorching tree bases. The temperature at the surface of the ground may be high, 90 to 120°C or even higher, but a few centimeters into the soil, it may not rise at all. Surface fires at intervals of years to tens of years were probably a natural occurrence in many kinds of vegetation in prehistoric times.

Ground fires may occur where there is a thick accumulation of litter. A ground fire is a flameless, subterranean fire that may burn slowly for long periods. It is hot and kills most of the plants rooted in the burning material. Even bogs can burn in this way because the heat from the ground fire dries out the peat ahead of it.

Crown fires occur in dense woody vegetation. Fire spreads rapidly through the canopy, killing the trees and most vegetation from the ground up. Usually crown fires seem to occur when surface fires are too infrequent to prevent the buildup of flammable material below the canopy.

It is true that humans start many fires but the overall result of modern man's activities has probably been a reduction in the amount of burning of vegetation. When fires start, men put them out, and man's land-use patterns have produced frequent firebreaks in such forms as roads, plowed fields, and asphalt parking lots. Indians and other primitive societies were reponsible for burning a great deal of the landscape. This seems to have been partially accidental in that Indians did not take fires very seriously and so took few pains to prevent them from spreading. There is evidence, however, that Indians deliberately set fires to drive game, and perhaps they knew that fires would favor certain useful plants or vegetation types. Fires started by lightning are so much a natural occurrence in many geographical areas that they should be considered an aspect of climate. It has been estimated that over 64% of wildfires in U.S. National Forests result from lightning.

The direct effects of surface fire on plants are fairly evident. Most herbs and shrubs are killed above ground but sprout from underground parts. Thick-barked trees, such as bur oak, tamarack, and ponderosa and longleaf pines, usually escape injury. Thin-barked trees such as black oak, white cedar, and lodgepole pine are usually killed; some thin-barked trees, however, have the ability to send up sprouts from their roots that allows them to survive fires in the same way as perennial herbs. Examples are black oak, beech, basswood and trembling aspen. Often the roots send up several sprouts so that a forest with trees having multiple trunks is a good indicator of a fire in the past.

Some species of plants are virtually dependent on fire to reproduce successfully. Certain pines, such as lodgepole and jack pine, have cones that stay on the tree and retain the seeds for many years unless the cones are heated or the tree is killed. When this happens the seeds are promptly shed with the resulting seedlings growing up to produce even-aged stands. For certain plants, such as the sandhill laurel and some prairie plants, exposing the seeds briefly to high temperature affects the seeds themselves, apparently breaking their dormancy and allowing them to germinate promptly. Such features would be advantageous only in environments in which fires are reasonably frequent; they are evidently adaptations to fire.

The indirect effects of fire are also important to the future of the plants and animals on the burned area. The amount of light is increased simply through the loss of some of the plant cover. As long as the surface remains blackened the soil tends to warm up more rapidly. There may be increased runoff of water and consequent erosion. The burning of litter frees inorganic nutrients such as calcium and potassium which resprouting and germinating plants can use; however, there may be some increased nutrient loss by leaching. With a light fire, the addition of bases to the soil tends to promote the growth of nitrifying bacteria and plants with high nitrogen requirements may be able to invade, such as fireweed, which often invades (or increases) in northern forests after fires. The removal of litter may allow the germination of seeds which germinate best on bare ground; the aspens are an example.

It seems clear that several aspects of the natural landscape of this country were dependent on fire. The pineries of both the Southeast and the Lake States are fire-based. In the absence of fire the pines give way to hardwoods. The prairies and bur oak savanna (called oak openings) of the Midwest were maintained by fire. Fires occurring every few years kept the bur oaks small, killed most other invading woody plants, and allowed the grasses and other herbs to grow vigorously the spring after a fire. In the West ponderosa pine and Douglas fir seem to be fire-dependent forests.

Chaparral vegetation evidently exists in a kind of continuous cycle in which fire is important. Areas of old shrubby growth build up a heavy, highly flammable litter of leaves, twigs, bark, and acorns. Fire removes the litter and thins the cover somewhat, allowing grasses and other herbs to flourish for a while until various effects of shrub regrowth cause them to decline. There is evidence in California chaparral that chemicals produced by the shrubs inhibit the herbs, and that one important effect of fire is to stop production of these chemicals and possibly to destroy those that have accumulated in the soil. Even in the climax hardwood forests of the eastern United States there is evidence that much of the species diversity is due to recurrent disturbance, particularly fire.

The scientific use of fire as a tool for ecological management originated in the southeastern U.S. through the work of such men as the Yale forester H. H. Chapman and the wildlife biologist Herbert Stoddard. Longleaf pine is a valuable timber tree, and the grasses and legumes that grow between the wide-spaced trees support good populations of bobwhite, a favorite game bird. Controlled burning to produce light surface fires eliminates the oaks and other hardwoods that would eventually dominate the area and also favors the grasses and legumes. Longleaf pine has a life history adapted to recurrent fires (Fig. 2–15). Early growth produces a short stem surrounded by dense needles; the young plant looks much like a clump of grass and is called the *wire-grass stage*. The plant shows little growth in height for 3 to 7 years; at this stage it is resistant to damage from fire because the terminal bud is buried in the middle of the dense clump of long, green needles, which burn very poorly. Although it doesn't get taller, the plant produces a deep, extensive root system, allowing it to grow rapidly at the end of the wire-grass stage. It may grow 4 to 6 feet a year for 2 or 3 years, carrying the terminal bud up out of the reach of surface fires. It is during this period of rapid growth that the plant is most vulnerable to fire. Soon afterwards the bark thickens enough to make the large sapling again resistant to fire.

Figure 2–15. Some aspects of adaptation to fire in longleaf pine. In the wiregrass stage (A), longleaf pine is resistant to surface fires because the bud is protected by a dense cluster of green, very long needles. This is in contrast to the seedling stage of most pines, such as slash pine of the same region, in which the terminal bud is much more exposed (B). The wiregrass stage may last three to seven years, during which the seedling grows little in height but instead puts its energy into food reserves and an extensive root system. These prepare the plant for rapid growth, with the tree shooting up eight to eighteen feet over a period of two to three years (C, D). This is the period of greatest vulnerability to fire but, by the end of it, the terminal bud is out of reach of surface fires and the bark has thickened enough to protect the trunk. Photographs are of longleaf pine flatwoods in Alachua County, Florida. Older longleaf pines are visible in the background.

Fire has also been recognized as a useful tool in such projects as providing the even-aged stands of jack-pine that the endangered Kirtland's warbler needs for its nesting habitat, and in preserving grasslands dominated by native prairie species. Burning in peatlands and boreal forests has been suggested as a means of increasing reindeer moss and thereby increasing woodland caribou populations.

Fire is, however, no cure-all. In the shortleaf pine forests of Kentucky, burning produces dense areas of root-sprouts from hardwoods that are worthless as timber and in which regeneration of shortleaf pine is impossible.

Pollution

Not long ago the word "pollution" generally referred to adding filth or poison to water or air but it has recently, and quite sensibly, been given a broader meaning. A good formal definition is that **pollution** is the unfavorable modification of the environment by man's activities.

Pollution is not new. Air pollution caused by burning coal was recognized as a human health problem by the mid-1600's, but pollution of various kinds was producing unrecognized problems long before that. Lead poisoning occurred among the upper classes in Ancient Rome as the result of lead-lined cookware and lead water pipes (our word "plumbing" comes from the Latin word for lead, *plumbum*). Even the scientific study of pollution goes back many years. Stephen A. Forbes and R. E. Richardson of the Illinois Natural History Survey studied the effects of sewage pollution on the Illinois River before 1920, and R. C. Osburn reported serious pollution of Lake Erie waters around the larger ports at the 1916 meeting of the Ecological Society of America.

Pollution may affect the individual organism directly by weakening or killing it but it also has effects at the population and ecosystem levels. One species may increase in numbers because pollution has harmed a species that competes with or preys upon it. Through such effects the community may change, often becoming less diverse. There is no absolutely satisfactory way to classify pollution, but some of the major categories that we will discuss in Chapter 6 are water, air, and thermal pollution, pesticides, and radioactive wastes.

ECOLOGICAL INDICATORS

"Every plant," wrote J. E. Weaver and F. E. Clements in their classic textbook in ecology, "is a product of the conditions under which it grows and is, therefore, a measure of environment." Although this is true, in practice we are interested in plants or animals (or their responses) that indicate some specific trait of its community or the habitat. We are interested in a species or a combination of species that tells us that the soil is salty, that a pasture is overgrazed, or that a stream is polluted. The ideal indicator, of course, would always be associated with the condition in which we are interested and would never occur in its absence.

The recent tendency to use ecological information for practical purposes is not new. The period from about 1945 to 1970 in which it was thought that chemical and physical technology would allow us to ignore ecological relationships was an aberration that it is hoped will not be repeated. Early in this century—and early in the development of ecology—the work of H. L. Shantz on plant communities of the Great Plains as indicators of crop capabilities was used for classifying public lands under the Stock Raising Homestead Act of 1916. Clements published a 400-page monograph on plant indicators in 1920, with most of it devoted to agricultural indicators such as indicators of crops and overgrazing.

As examples of indicator species, heavy grazing in the dry prairies of Wisconsin used as pasture is shown by a decrease in little bluestem, an increase in annual ragweed, and the invasion of dandelion and white clover. Other indications of overgrazing, in the Midwest or elsewhere, are an increase or invasion of spiny plants such as thistles and prickly pear cactus, and also an increase in unpalatable species.

There has been a great deal of interest in biological indicators of pollution. Portions of a stream dominated by certain annelid worms (*Tubifex* and *Limnodrilus*) or a kind of midge fly larva are generally heavily polluted by organic wastes such as sewage. If certain gill-breathing mayflies, stoneflies, and caddisflies are present the water is clean. Subdivisions in between these can also be recognized.*

Ecological indicators are also used to determine the depth of ground water, map soil types and glacial deposits, and determine types of rock formations, including prospecting for uranium. The applications of ecological indicators do not have to be practical, of course. By the use of evidence of fire, of former cultivation, and the like, we can often reconstruct a great deal of the history of an area. There is a fascinating book, *Reading the Landscape of America* by M. T. Watts, devoted to this subject.**

DISPERSAL AND RANGE EXPANSION

T. T. Macan, an aquatic ecologist, recently formalized a view of the distribution of organisms that goes back at least to the 1920's in the writings of David Starr Jordan (also an aquatic biologist who was president of Indiana University and later Stanford. In those days of smaller university enrollments Jordan is said to have prided himself on learning the name of every student on campus until he discovered that when he learned a new student he forgot the name of a fish). Macan has suggested that species will occupy an area if (1) the area provides their habitat requirements (both proximate and ultimate); (2) it is not made unsuitable by competitors, predators, or disease; and (3) the species can get there.

---

*Another approach to the problem of a pollution indicator has been to use a mathematical index based on the number of species and their relative abundances. This is discussed in a later section on the Shannon-Wiener index in Chapter 4.
**New York, Macmillan, 1975.

We have discussed (1) and (2) and now deal with getting to the area, or dispersal.

A formal definition of **dispersal** is the movement of individuals from the homesite. The means of dispersal vary from species to species. Among plants there is often a special life history stage (seeds or spores) having special structural features that seem to be adaptations for achieving dispersal. Examples are the winged fruits of maples or the plumed fruits of dandelion that use the wind and the stick-tight fruits of beggar's-ticks and spiny fruits of cocklebur that use animals. Structural adaptions for dispersal are not evident in many animals, probably because most animals can move about readily throughout their lives.

It is almost always the immature organism that does the dispersing. This is obviously true of plants and of sessile marine animals such as barnacles (in which dispersal is by the larvae), but it is also true of many other animals. In birds the adult is clearly just as motile as the young but it is the young that moves. In his classic study of the house wren S. C. Kendeigh found that 84% of the adult males and 70% of the adult females that survived the winter bred within 1000 feet of their nest of the previous year. Of the young that lived through the winter, however, only 15% nested within 1000 feet of where they had hatched. The adults tended to return each year to the site where they had settled in their first breeding season, a tendency called *site tenacity* (among several other names, including *philopatry* and *ortstreue*). The young tended to move from the immediate vicinity where they were hatched or born but stayed in the general area (of the house wren young, 70% nested in the zone between 1000 and 10,000 feet of where they had hatched).

The general pattern of dispersal is that most of the seeds or young animals, or whatever is doing the dispersing, settle close to the homesite. There is then a rapid decline with distance, and only a few individuals reach very substantial distances. If the number reaching or going beyond a certain distance is plotted against distance, the resulting curve looks like that in Figure 2–16.

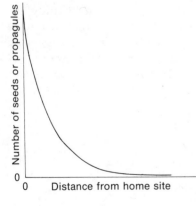

Figure 2–16. A common pattern of dispersal for organisms, especially those whose dispersal is passive—seeds with no special dispersal devices, for example. The decline with distance seems to result from the loss of seeds, either by lodging somewhere earlier or being eaten or otherwise destroyed, and from the increasingly larger area involved with every increase in distance from the center.

There are basically two ways of increasing dispersal—that is, of getting more individuals further from the homesite. One is increasing the number of seeds or other dispersing units. This is usually expensive (in the currency of energy) and not very effective. Doubling the number of seeds simply doubles the number reaching any given distance. If on the average 0.00001 seed is able to disperse a mile from the plant where it was produced, then if the plant produces twice as many seeds, 0.00002, on the average, will travel a mile.

The other method of improving dispersal is to increase *dispersal ability*. For plants, as already mentioned, this seems to have taken the form of structural adaptations using wind, water, or animals as dispersal agents. For animals dispersal ability seems to be improved by behavioral means which may include adult intolerance to young, forcing them to keep moving, and perhaps some greater tendency for the young to wander.

Most dispersal does not produce any change in the geographical range of the species. The dispersal is within the range already occupied, perhaps into a newly created area of suitable habitat (dandelions into a new lawn) or just into a spot left vacant when an older individual died. Successful dispersal into an area not formerly occupied by the species is called *range expansion.* The usual events allowing range expansion are (1) a barrier to the species is somehow removed or circumvented; or (2) a formerly unsuitable area becomes suitable. A third possibility in which evolutionary change in a species allows it to make use of a formerly unsuitable area almost certainly occurs but is hard to study.

Removal or circumvention of barriers usually occurs by one of two agents: geological processes, such as the joining and separating of continents; and human activities. The results of the removal of barriers through geological processes are difficult to study except by means of fossils. We know, for instance, that North and South America did not have a land connection such as the present Central American isthmus through much of the preceding 70 million years. During this period the two continents had almost no families of mammals in common. By glacial times, after the establishment of a connection between the two continents, there was rapid invasion in both directions. At the present time about 14 families of mammals are found in both North and South America.

There are many examples of human removal or circumvention of barriers that have occurred in the lifetime of ourselves or our parents. Humans brought starlings, house sparrows, and Dutch elm disease across the Atlantic Ocean. Human activities have reached the magnitude of geological events. The Welland Canal was dug (1914-1932) between Lakes Ontario and Erie, bypassing Niagara Falls which had been a barrier to the westward spread of the sea lamprey. As a result the lamprey reached the Detroit River in the 1930's and by the 1950's had spread all the way to western Lake Superior. A proposed sea level canal across Panama (or somewhere else in Central America) is a similar project. Such a canal would allow the marine biota of the Pacific to spread into the Atlantic, and vice versa (the present Panama Canal does not do this because ships are locked up about 25 meters above sea level and through two fresh water lakes for the crossing). The devastating effect of the introduction of the sea lamprey on the Great Lakes trout population is well known; the results of range expansions following construction of a sea level canal across Central America can only be speculated on.

In the category of environmental changes allowing range expansion, changes in climate and vegetation seem most important. The clearing of the land in the eastern and midwestern U.S. allowed the westward spread of several grassland birds, including the horned lark and the Savannah sparrow. Several species of birds and mammals have extended their ranges northward in the past 50 to 75 years, possibly in response to a climatic warming trend (Fig. 2–17); one example is the cardinal. The cardinal was absent from Michigan (and comparable areas in Wisconsin and westward and in Ontario and eastward) in the late 1800's. The first known nesting in Michigan was in 1892. By 1920 it occupied a range about one third of the way up the state; by 1950 it had reached the base of the "thumb" and in a rapid expansion over the next 20 years it had reached the northern tip of the lower peninsula.

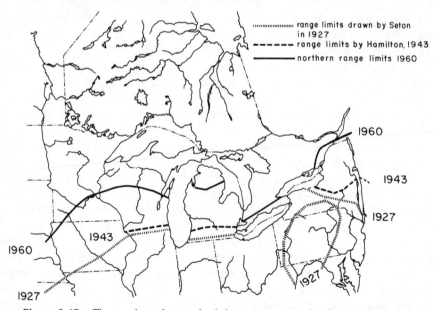

Figure 2–17. The northward spread of the opossum in the Great Lakes region. A hundred years ago the opposum was absent from most of the northern parts of Iowa, Indiana, Ohio, and Pennsylvania. The 1927 line shown on the map is the northern range boundary drawn by Ernest Thompson Seton, and the 1943 line is the northern range boundary in W. J. Hamilton's *Mammals of the Eastern United States*. (From A. De Vos, "Recent changes of the ranges of mammals in the Great Lakes region," *American Midl. Nat.*, 71:222, 1964.)

Similar patterns are known from Europe. A hundred years ago the nightingale in Scandinavia occurred only in southern Sweden. Since then it has spread as far north as Stockholm and Uppsala. Because Europe has a much longer historical record than North America, however, we know that the nightingale was common around Stockholm and Uppsala at the time of Linnaeus in the middle 1700's. It had retreated southward in the next hundred years and then readvanced more recently. This points up the fact that geographic ranges are dynamic. Boundaries fluctuate in response to many environmental factors and the shifts, especially for animals and short-lived plants, may be rapid.

BIBLIOGRAPHY

Aron, W. I., and Smith, S. H. "Ship canals and aquatic ecosystems," *Science*, 174:13–20, 1971.

Baker, J. R. "The evolution of breeding seasons," *in* G. R. deBeer, ed., *Evolution: Essays on Aspects of Evolutionary Biology*. New York, Oxford University Press, 1938.

Berndt, R., and Winkel W. "Gibt es beim Trauerschnäpper *Ficedula hypoleuca* eine Prägung auf den Biotop des Geburtsortes?" *J. Ornithol*, 116:195–201, 1975.

Black C. A. *Soil-Plant Relationships*, 2nd ed. New York, John Wiley & Sons, 1968.

Black, C. C. "Ecological implications of dividing plants into groups with distinct photosynthetic production capacities," *in* J. B. Cragg, ed., *Advances in Ecological Research*, Vol. 7. New York, Academic Press, pp. 87–114, 1971.

Blackman, F. F. "Optima and limiting factors," *Ann. Botany*, 19:281–295, 1905.

Braun-Blanquet, J. *Plant Sociology: The Study of Plant Communities*, (transl., rev., and ed. by G. D. Fuller and H. C. Conard). New York, McGraw-Hill, 1932.

Brock, T. C. "High temperature systems," *Ann. Rev. Ecol. System.*, 1:191–220, 1970.

Brody, Samuel. *Bioenergetics and Growth*. New York, Reinhold, 1945.

Bünning, E. *The Physiological Clock* (revised 2nd ed.). Berlin, Springer-Verlag, 1967.

Clements, Frederick E. *Plant Succession and Indicators*. New York, H. W. Wilson, 1928.

Daubenmire, R. F. "Ecology of fire in grasslands," *in* J. B. Cragg, ed., *Advances in Ecological Research*. Vol. 5. New York, Academic Press, 1968, pp. 209–266.

Davis, E. A., Jr. "Seasonal changes in the energy balance of the English sparrow," *Auk*, 72:385–411, 1955.

Dawson, W. R. and Schmidt-Nielsen, K. "Terrestrial animals in dry heat: desert birds," *in* D. B. Dill, ed., *Handbook of Physiology*, Sect. 4, Washington, American Physiological Society, 1964, pp. 481–492. Also several other articles by other authors in the same source.

Eastman, J. A. "In wildness is fire," *Living Wilderness*, 36(117):11–17, 1972.

Eyre, S. R. *Vegetation and Soils*. Chicago, Aldine, 1963.

Gates, David M. *Energy Exchange in the Biosphere*. New York, Harper & Row, 1962.

———. "Transpiration and leaf temperature," *Ann. Rev. Plant Physiol.*, 19:211–239, 1968.

Geiger, R. *The Climate Near the Ground*, 4th ed. Cambridge, Mass. Harvard University Press, 1965.

Gessaman, J. A. "Ecological energetics of homeotherms," *Utah State Univ. Press Monogr. Series*, no. 20, pp. 1–155, 1973.

Hildén, O. "Habitat selection in birds: a review, "*Ann. Zool. Fennici*, 2:53–75, 1965.

Jaeger, E. C. "Further observations on the hibernation of the poor-will," *Condor*, 51:105–109, 1949.

Jordan, David Starr. "The distribution of fresh-water fishes," *Ann. Rep. Smithsonian Inst. for 1927*, pp. 355–385, 1928.

Kendeigh, S. Charles. "The role of environment in the life of birds," *Ecol. Monogr.*, 4:299–417, 1934.

———. 1941. "Territorial and mating behavior of the house wren," *Ill. Biol. Monogr.*, 18:2–120, 1941.

———— and J. Pinowski, eds. *Productivity, Population Dynamics, and Systematics of Granivorous Birds.* Warsaw, Institute of Ecology, Polish Academy of Science, 1973.

Kennedy, J. S. "Host-finding and host-alternation in aphids," *Proc. 8th Int. Congr. Entomol.* (Stockholm) 1948:423–426, 1950.

Leopold, A. S., Erwin, M., Oh, J., and Browning, B. "Phytoestrogens: adverse effects on reproduction in California quail," *Science,* 191:98–100, 1976.

Macan, T. T. *Freshwater Ecology,* 2nd ed. New York, John Wiley & Sons, 1974.

Maximov, N. A. "The physiological significance of the xeromorphic structure of plants," *J. Ecol.,* 19:273–282, 1931.

Morehouse, E. L., and Brewer, R. "Feeding of nestling and fledgling Eastern kingbirds," *Auk,* 85:44–54, 1968.

Muller, C. H., Hanawalt, R. B., and McPherson, J. K. "Allelopathic control of herb growth in the fire cycle of the California chaparral," *Bull. Torrey Bot. Club,* 95:225–231, 1968.

Proctor, J., and Woodell, S. R. J. "The ecology of serpentine soils," *in* J. B. Cragg, ed., *Advances in Ecological Research,* Vol. 9. New York, Academic Press, 1975, pp. 255–366.

Prosser, C. L., *Comparative Animal Physiology,* 3rd ed. Philadelphia, W. B. Saunders, 1973.

Raunkiaer, C. *The Life Forms of Plants and Statistical Plant Geography.* Oxford, Clarendon Press, 1934.

Richardson, R. E. "Changes in the bottom and shore fauna of the middle Illinois River and its connecting lakes since 1913–1915 as a result of increase southward of sewage pollution," *Ill. St. Nat. Hist. Surv. Bull.,* 14:33–75, 1921.

Schindler, D. W. "Evolution of phosphorus limitation in lakes," *Science,* 195:260–262, 1977.

Schmidt-Nielsen, Knut. *Desert Animals. Physiological Problems of Heat and Water.* London, Oxford University Press, 1964.

Shantz, H. L. "Natural vegetation as an indicator of the capabilities of land for crop production in the great plains area," U.S. Dept. Agr., Bureau of Plant Industry. Bull. No. 201, 1911.

Svärdson, G. "Competition and habitat selection in birds," *Oikos,* 1:157–174, 1949.

Taylor, Walter P. "Significance of extreme or intermittent conditions in distribution of species and management of natural resources, with a restatement of Liebig's law of the minimum," *Ecology,* 15:374–379, 1934.

Tinbergen, N. *The Study of Instinct.* New York, Oxford University Press, 1951.

Turreson, G. "The genotypical response of the plant species to habitat," *Hereditas,* 3:211–350, 1922.

Twomey, Arthur C. "Climographic studies of certain introduced and migratory birds," *Ecology,* 17:122–132, 1936.

Udvardy, Miklos D. F. *Dynamic Zoogeography.* New York, Van Nostrand Reinhold, 1969.

Vance, B. D., and Kucera, C. L. "Flowering variations in *Eupatorium rugosum,*" *Ecology,* 41:340–345, 1960.

Vander Veen, R., and Meijer, G. *Light and Plant Growth.* Philips, 1959.

Vernberg, F. J., and Vernberg, W. B. *The Animal and The Environment.* New York, Holt, Rinehart, & Winston, 1970.

Weiser, C. J. "Cold resistance and injury in woody plants," *Science,* 169:1269–1278, 1970.

Wilhm, J. F. "Biological indicators of pollution," *in* B. A. Whitton, ed., *River Ecology.* University of California Press, Studies in Ecology, vol. 2, 1975.

Williamson, M. J. "Burning does not control young hardwoods on shortleaf pine sites in the Cumberland plateau," U.S. Forest Service Res. Note CS-19:1–4, 1964.

Chapter 3

POPULATION ECOLOGY

Populations have traits of their own that differ from those of the individuals composing the populations. An individual is born once and dies once, but a population continues, perhaps changing in size depending on the birth and death rates of the populations. An individual is male or female, old or young, but a population has a sex ratio and an age structure. This chapter examines some of the ecological traits unique to the population level of integration.

BIRTH AND DEATH

BIRTH RATE AND DEATH RATE

New individuals can be added to populations in two ways, by birth and by immigration. Individuals can leave populations in two ways, death and emigration. Generally births and deaths are expressed as rates, that is, as numbers in a given time. If there are 120 births in a population during a year, the birth rate (*natality rate*) is 120 per year or 10 per month. We can express death rate (*mortality rate*) in the same way.

If we ignore immigration and emigration—which is easy enough to do on paper but generally impossible in nature—then changes in population size depend on the balance between birth and death rates. If the two are equal then the population size is stable. If birth rate exceeds death rate the population grows and if death rate exceeds birth rate the population declines.

LIFE TABLES AND LONGEVITY

A *life table* summarizes the statistics of death and survival of a population, by age. A made-up life table, just to illustrate the terms and calculations used, is shown in Table 3–1 for the McKinley murre population on Mount Deevey. Ages are in years but for other kinds of organisms they could be days, months, or hours. The age at the beginning of

Table 3–1   LIFE TABLE FOR McKINLEY MURRE POPULATION

Age, x	Survivorship, l_x	Mortality, d_x	Mortality Rate, q_x	Life Expectation, e_x
0	100	55	0.55	1.15
1	45	30	0.67	0.94
2	15	10	0.67	0.83
3	5	5	1.00	0.50
4	0	—	—	—

the interval is represented by x; l_x is survivorship, the number of individuals alive at the beginning of the interval; d_x is mortality, the number dying within the interval; q_x is mortality rate, the number dying within the interval divided by the number alive at the beginning of the interval; and e_x is life expectation, the average time left to an individual at the beginning of the interval.

In this table we start out with 100 individuals at the time they are born (age 0). Such a group, all born at the same time, is called a *cohort;* all the male children born in the United States in June 1955 would be a cohort. All are alive at the beginning of the interval but in the table over half (55) died during the first interval. For the murres in their first year, then, mortality rate is 55% (55/100 × 100). A murre at birth can expect to live, on the average, just over 1 year. Since 55 animals died in the first year, 45 survived to begin the second. During this period, from age 1 to 2, 30 died. Mortality rate in the second year, in this case, is slightly higher than for the first year, 67% (30/45 × 100). The murre aged 1 year has, on the average, slightly less than 1 year of life remaining to him.

About the only part of such a table which is not easy to understand is the calculation of life expectation. The concept itself is simple enough; it is simply the average additional time still left to individuals alive at a particular age. The life expectation for a white male in the U.S. as shown in Table 3–2 is 68.3 years when he is born. When he is 1 year old his life

Table 3–2   LIFE TABLES FOR WHITE MALE AND FEMALE HUMANS IN THE UNITED STATES, AS OF 1972

Women have considerably better survivorship so that the mean expectation of further life at birth is more than seven years longer for females than for males. Notice, though, how a life table such as this one is constructed. It does not really start out with 100,000 children born in 1972; we could not get their vital statistics until well into the twenty-first century. Instead, it uses current death rates for each age group. The mortality rate for young children is for persons born in the 1970's, but the mortality rate for the oldest age categories is for persons born in the latter part of the nineteenth century. The other columns are all calculated from the mortality rate column. Therefore we cannot really say that a girl born now will live longer, on the average, than a boy. It may be so, but it also may be that social conditions will change enough that women will suffer increasingly from the stress-related diseases such as heart disease and high blood pressure that now carry off men at early ages. Life tables cannot predict the future or, rather, they predict the future only when conditions do not change.

Table 3-2 LIFE TABLES FOR WHITE MALE AND FEMALE HUMANS IN THE UNITED STATES, AS OF 1972*

Age, x (years)	Mortality Rate, q_x	Survivorship, l_x	Mortality, d_x	Life Expectation, e_x
White, Male				
0-1	0.0182	100,000	1,824	68.3
1-5	.0033	98,176	323	68.5
5-10	.0023	97,853	220	64.7
10-15	.0024	97,633	238	59.9
15-20	.0075	97,395	735	55.0
20-25	.0096	96,660	927	50.4
25-30	.0083	95,733	799	45.9
30-35	.0089	94,934	845	41.3
35-40	.0123	94,089	1,154	36.6
40-45	.0198	92,935	1,840	32.0
45-50	.0331	91,095	3,013	27.6
50-55	.0516	88,082	4,542	23.5
55-60	.0827	83,540	6,913	19.6
60-65	.1255	76,627	9,618	16.1
65-70	.1793	67,009	12,014	13.1
70-75	.2575	54,995	14,160	10.4
75-80	.3624	40,835	14,798	8.1
80-85	.4745	26,037	12,354	6.3
85 and over	1.0000	13,683	13,683	4.7
White, Female				
0-1	0.0137	100,000	1,370	75.9
1-5	.0025	98,630	252	75.9
5-10	.0017	98,378	164	72.1
10-15	.0014	98,214	141	67.2
15-20	.0029	98,073	285	62.3
20-25	.0032	97,788	314	57.5
25-30	.0035	97,474	339	52.7
30-35	.0047	97,135	461	47.9
35-40	.0071	96,674	687	43.1
40-45	.0114	95,987	1,093	38.4
45-50	.0180	94,894	1,710	33.8
50-55	.0260	93,184	2,425	29.3
55-60	.0402	90,759	3,649	25.1
60-65	.0585	87,110	5,096	21.0
65-70	.0899	82,014	7,370	17.1
70-75	.1445	74,644	10,784	13.6
75-80	.2361	63,860	15,078	10.4
80-85	.3534	48,782	17,237	7.8
85 and over	1.0000	31,545	31,545	5.7

*From *Vital Statistics of the United States 1972*, vol. 2, part A, Table 5-1, U.S. Department of Health, Education, and Welfare, Public Health Service.

expectation is not 67.3 years but 68.5, a little more than at birth. This is because the death rate for children under 1 year old is high, higher than at any other time until after the age of 40. Similarly, the life expectation of a 68-year-old man is not 0. Most who make it to 68 live a while longer—about 12 years longer, on the average.

Life expectation at age 0 is the same thing as *mean natural longevity.* Another aspect of longevity is *physiological longevity,* the age reached by individuals dying of old age—individuals living under conditions where such causes of death as predation, accident, poor nutrition, and infectious disease are not factors. Although this may seem to be an ill-defined concept, it is nevertheless useful. There are upper limits to the ages of most kinds of organisms, a time of life when senility takes over no matter how good the conditions of life are. There are no 2000-year-old men, nor even 200-year-old ones, just as there are no 100-year-old horses nor 50-year-old dogs.

Plotting the l_x (survivorship) column of a life table against the x (age) column gives a *survivorship curve* (Fig. 3–1). This is convenient for use as a visual aid to detect changes in survivorship (and mortality) by period of life.

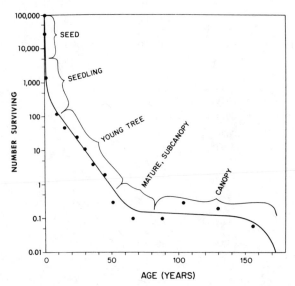

Figure 3–1. A survivorship curve for a tree. This particular tree is a tropical cabbage palm *(Euterpe globosa)* that grows in the upland forests of Puerto Rico. Survivorship curves are usually plotted logarithmically on the y-axis, as shown here, for two reasons. A very tall graph is avoided and, more importantly, a straight portion of the survivorship curve indicates a period when the mortality rate is not changing. During the young tree period shown on this curve the mortality rate is near constant, about 10 to 15% per year. It then drops (and survival rate rises) as the tree matures. Then, somewhere around 150 years of age, the mortality rate increases greatly. (From L. Van Valen, "Life, death, and energy of a tree," *Biotropica,* 7:263, 1975.)

POPULATION GROWTH

EXPONENTIAL POPULATION GROWTH

When a very small population is living in a very large area of favorable habitat, population growth rate tends to depend on two factors: the size of the population and the capacity of the population to increase (in other words, the *biotic potential*). A situation of this sort occurs for aquatic organisms when a new lake is formed, whether by glaciers or the Corps of Engineers. For deer it may occur when they somehow manage to reach an island where they have never lived before. In the laboratory we can readily create such a situation by mixing up a bottle of culture medium and adding some microorganism to it. Let us examine such a case. For the sake of simplicity, we will use an organism that splits into two and assume that under thoroughly favorable conditions this occurs every 4 hours.

We will start at time zero with one such animal. Four hours later it will divide and we will have two. By the end of 8 hours, these two will divide and there will be four swimming in the culture medium. Four hours later, after half a day, there will be eight, and after another 12 hours, at the end of the first day, there will be 64 organisms from the original one. Four hours into the second day there will be 128 organisms and by the end of the day there will be 4096. By the end of the third day there will be 262,144 organisms. By the end of the fourth day there will be 16,777,216, which is not so many, but by the end of the fifth day there will be 1,073,741,824.

This type of growth, in which the curve of numbers vs. time becomes steeper and steeper, is called *exponential growth* (Fig. 3–2). Notice that

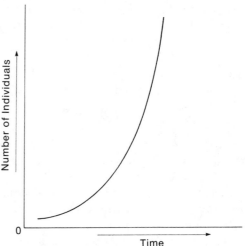

Figure 3–2. An Exponential growth curve. In exponential population growth, growth rate depends on the biotic potential, which is a constant, and the number of reproducing individuals, which is steadily increasing. Both growth rate and the total population size (plotted here) rise at a steadily increasing rate.

the growth depends on the biotic potential, but this does not change; a division occurs every 4 hours all through the 5 days. Growth also depends on the size of the population and this, of course, does change, continually growing larger. As a consequence, the growth rate of the population steadily increases from a slow rate, when the population is low (an increase of only 63 animals in the first day) to a fast rate when the population is high (an increase of over 1 billion animals by the fifth day).

BIOTIC POTENTIAL

The *biotic potential,* we have said, is the capacity of a population for increase. It is often represented by the symbol r (little "r"). Another name for biotic potential is *intrinsic rate of natural increase,* a term coined by the insurance actuary A. J. Lotka, who was a pioneer in population theory. Any thorough discussion of biotic potential quickly becomes mathematical; here we will say only that it is a population increase rate and is determined by both birth and death rates. More specifically, it is the population growth rate per head (per individual member of the population) when the population is uncrowded and has a particular type of age distribution (the stable age distribution; see p. 81). Note that the population will actually grow at this rate only when it is uncrowded; much of the time the biotic potential is, in fact, potential and not actual. Nevertheless, if two populations, one with a high r and one with a low r, each suffers some catastrophe that drastically lowers its numbers, the species with the high r will be able to rebound more rapidly.

Some of the traits that influence whether the biotic potential of a population is high or low are the following:

1) Number of offspring per breeding period (for example, number of young in a litter, number of eggs in a clutch)
2) Survival up to and through reproductive age
3) How long the reproductive age of the organism lasts (at one extreme are organisms such as salmon that breed only once; many other types of organisms breed every year until death or old age)
4) Age at which reproduction begins

Obviously if an organism has a large litter, a high survival rate, an early breeding age, and reproduces repeatedly, it will have a higher r than one which has small litters, low survival, does not mature until older, and breeds only once. However, it is not easy to see which of these factors tend to be relatively more important without some mathematics. Which of these traits, if altered, will most affect biotic potential? The Cornell University ecologist LaMont C. Cole showed that age at first breeding is exceedingly influential in determining whether biotic potential is high or low (Fig. 3–3). This may be surprising at first but a little reflection will show the logic of it. Imagine two species which are similar except that one begins to breed at one year of age and the other at three years. At the end of the second year, the first species will have daughters who will be breeding and producing daughters who will, themselves, be breeding and producing more daughters in the third year, when the second species is just beginning to breed.

In suboptimal environments, of course, even species with life history traits favoring rapid, prolific reproduction will have a low r because of abiotic factors which lower birth rate, raise death rate, or slow development. In fact, looking at the geographical range of a species in terms of population ecology, the range boundaries are approximately the line at which r is zero. Outside the line populations of the species, if they somehow get a start, will decline, while inside the line the populations will have a positive capacity for growth (Fig. 3–4). The relationship of the biotic potential of a grain beetle to temperature and moisture is shown in Figure 3–5.

The ranges of the intrinsic rate of natural increase for various kinds

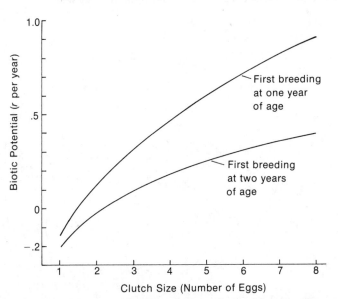

Figure 3–3. The influence of number of eggs and age of first breeding on biotic potential, or intrinsic rate of natural increase, in birds. Biotic potential (r per year) is plotted against clutch size (number of eggs laid for one nesting). Calculations are based on life-history data intended to represent a small- or medium-sized open-nesting bird such as a song sparrow or an American robin. Increasing clutch size increases r; delaying maturity decreases r. A bird which does not breed until it is two years old but then lays eight eggs every year has a lower r than one that lays only four eggs per year but begins breeding when it is one year old. (From R. Brewer and L. Swander, "Life history factors affecting the intrinsic rate of natural increase of birds of the deciduous forest biome," *Wilson Bulletin,* 89:218, 1977.)

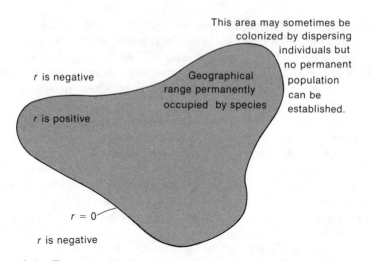

Figure 3–4. The geographical range of a species can be thought of as the region within which the intrinsic rate of natural increase (*r*) of its populations is greater than zero. Although this is generally true, it ignores a few complications. The geographical range of long-lived plants may, for example, include areas in which, because of climatic changes, they can no longer reproduce but where they can continue to survive very well.

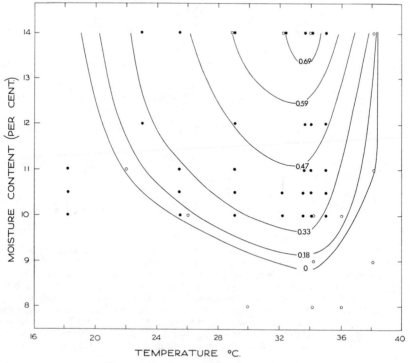

Figure 3–5. See opposite page for legend.

Table 3–3 INTRINSIC RATE OF NATURAL INCREASE

Kind of Organism	Approximate Biotic Potential, r (per year)
Large mammals	0.02–0.5
Birds	0.05–1.5
Small mammals	0.3–8
Larger invertebrates	10–30
Insects	4–50
Small invertebrates (including large protozoans)	30–800
Protozoa and unicellular algae	600–2000
Bacteria	3000–20000

of organisms are shown in Table 3–3. Because the exact determination of biotic potential is tedious, time-consuming, or difficult for most kinds of organisms, the values are based on a few good studies plus several estimates. These ranges probably include the bulk of the species in each group, but there are exceptions. For example, the 17-year cicada is older when it reaches breeding age than elephants, rhinoceroses, or humans; it has a biotic potential somewhere around 0.4 per year, far below that of most insects.

LOGISTIC POPULATION GROWTH

Calculations such as those demonstrating exponential population growth for microorganisms always seem faintly ridiculous because the numbers of most organisms we are acquainted with usually stay about the same from year to year (man is a temporary exception). If we have a pair of house wrens nesting in our yard this year we know that the likelihood is small that we will have 4 pairs next year, 16 the next year, and so on. At some point in the growth of a population, then, growth must slow down and tend toward a zero growth rate.

Figure 3–5. The relationship of the intrinsic rate of natural increase (given as r per week) of a grain beetle to temperature and moisture. Under a specified set of environmental factors r is a constant, being the instantaneous growth rate per head when the population is uncrowded and has a stable age distribution. Between habitats or geographical regions, however, r may vary, as birth rate, death rate, or speed of development are changed. The figure is similar to a topographic map which shows elevations by contour lines, with temperature on one axis and moisture on the other. The value obtained for r is plotted at the intersection of lines for each combination of temperature and moisture. Lines are then drawn which enclose equal values of r. The highest values of r, corresponding to the top of a hill on a topographic map, are in the area at about 34°C and 14% moisture. Moving away from this peak, r decreases (more rapidly approaching a higher temperature than a lower one) until it equals zero. Beyond this point the population could not grow but could only decline, so that habitats having combinations of temperature and moisture in this region are not permanently inhabitable by the species. (From L. C. Birch, "Experimental background . . . I. The influence of temperature, moisture, and food on the innate capacity for increase . . . ," *Ecology*, 34:707, 1953.)

When the growth curve of many populations is drawn—when population size is plotted against time—the curve that results looks like a flat S and is called a *sigmoid growth curve* (Figs. 3–6 and 3–7). In this curve it seems as though population growth is exponential or approximately so at the beginning; the growth rate starts out slowly and then gets faster and faster. Then, when the population is medium-sized, the growth rate begins to slow down and becomes slower and slower until it finally becomes zero when births balance deaths. The simplest model for describing this type of growth is an equation called the *logistic* equation, introduced to ecology by Raymond Pearl and L. J. Reed. What it says, in effect, is that growth rate of the population is determined by the biotic potential

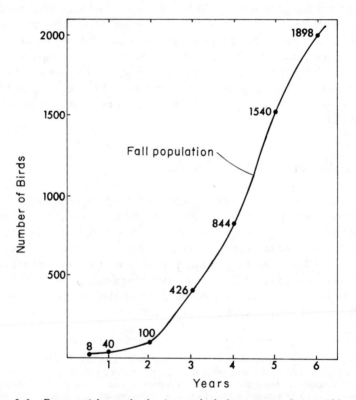

Figure 3–6. Exponential growth of a ring-necked pheasant population. Although there are many examples of exponential growth in laboratory studies, there are not many field examples because few have taken the trouble to obtain accurate censuses. In this study, two cocks and eight hen pheasants were released on Protection Island off the coast of Washington. Two of the hens died immediately but reproduction by the rest brought the population up to 40 by the end of the first breeding season. Five years later there were 1898 pheasants on the 200 acres of suitable habitat on the island. Early growth was exponential but growth rate dropped in the last years of the study. Would the curve have levelled off smoothly or would it have overshot? In the fall of 1943, the 1898 had decreased to 732 as a result of several occurrences that ended the experiment: 450 birds were trapped and removed by the Washington Game Department for restocking elsewhere; disturbance from nighttime trapping caused many birds to fly out over the ocean and drown; and the food supply was decreased since 100 head of cattle had been moved onto the island for the winter. (From A. S. Einarsen, "Some factors affecting ring-neck pheasant density," *Murrelet*, 26:7, 1945.)

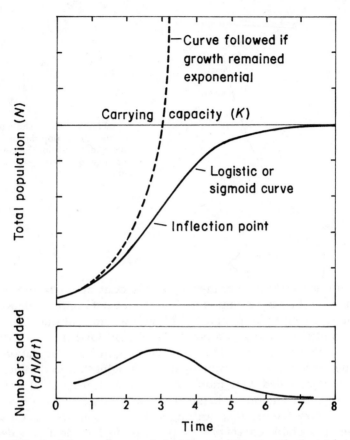

Figure 3–7. Logistic growth curve. In logistic population growth, the total size of the population grows in an S shape, first growing slowly, then faster, then more slowly, levelling off near the carrying capacity. Growth rate (plotted in the lower graph) increases, reaches a peak in the middle of the curve of numbers (the point of inflection, where the curve stops bending to the left and starts bending to the right), and then decreases toward zero as population numbers approach the carrying capacity.

and the size of the population as modified by the *environmental resistance* or, in other words, by all the various effects of crowding. These effects may include lowered reproduction because of poor nutrition of the mother, high death rates because of predators, and increased emigration, among other things. Environmental resistance increases as population size gets closer to the *carrying capacity* (usually represented by K), which is the population size that the area has the resources to support.

Not all populations show a growth curve such as we have just described. Population growth very frequently seems to overshoot the carrying capacity and then drop back rather sharply, so that the first part of the curve looks something like a J (Fig. 3–8). What causes such a curve as this? One likely reason is that there is a lag between the time at which the population attains a certain size and the time at which the unfavorable effects of that level of crowding are felt; in the meantime, the population continues to grow. What could cause such a time lag? If predators are an

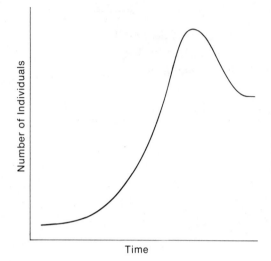

Figure 3–8. J-shaped growth curve. Often population numbers seem to overshoot the carrying capacity and then drop back to (or below) it.

important aspect of environmental resistance it might take a while for them to reproduce or to immigrate into the area of high population. Until their numbers build up population growth would continue. Or, suppose that the nutrient supply in a pond is sufficient for 5 million individuals of some organism that reproduces by splitting in two. Population growth continues until there are 4 million, at which point the environment is still relatively uncrowded. One more division of each individual would raise the number to 8 million, well above the carrying capacity. A die-off to or possibly below the original carrying capacity would follow. Or suppose that the unfavorable effects of high populations include a slow decline in soil fertility and a slow buildup of poisonous materials produced by the organisms, effects which are not serious until they reach a certain level, a threshold, at which point they become very serious. If you think that this last example sounds as though it might have something to do with human populations, you are not alone.

Remember that the logistic equation is just one mathematical way of representing a form of population growth. In other words, it is a model of what the organisms may be doing; it is not a law they are following. As a model it is oversimplified and in some respects it is certainly wrong for some organisms. A few of the difficulties with it are these:

(1) It assumes a straight-line relationship between density and the unfavorable effects of crowding, which is probably sometimes the case and sometimes not. There are, in fact, indications for some kinds of organisms that at very low densities *undercrowding* is also unfavorable. (This is known as the *Allee effect,* named for W. C. Allee, who described it in his book *Animal Aggregations.**)

---

*Chicago, University of Chicago Press, 1931.

(2) The simple logistic does not allow for time lags; everything is instantaneous. This is clearly not true in nature where, as we already mentioned, predators may respond to a crowded prey population by immigrating, a process that may take days or weeks.

(3) The logistic equation assumes that every individual in the population is the same. This may be valid for bacteria but in a population of mammals there will be young and possibly old animals who will not be reproducing at all, and females of different ages who may well have different-sized litters.

If the logistic model has these and other defects, why use it? Because (in addition to its being mathematically easy) it is still an adequate representation of population growth for many purposes. In similar fashion, your car may not be in perfect condition, but if it gets you where you want to go you may hesitate to discard it.

MATHEMATICAL REPRESENTATIONS OF POPULATION GROWTH

The formula by which exponential population growth is generally represented is

$$dN/dt = rN$$

This is a differential equation in which dN/dt is the population growth rate. It refers to the change in numbers (dN) per time interval (dt) when this interval is very small, so that we are taking an instantaneous reading of rate of growth. (A speedometer gives us such an instantaneous reading of our rate of travel in a car but regrettably there is no "growthometer" to attach to populations.) Biotic potential, r, is the increase in numbers of individuals per time period per head (that is, per individual). It combines birth rate and death rate, being equal to the instantaneous birth rate minus the instantaneous death rate. N is the number of individuals in the population.

Growth rate is higher in a population with a high r compared with one with a low r. For example, if two populations each have a population size of 100, and r for one is 0.5 and for the other 0.1, then dN/dt for the first population will be 100 × 0.5 or 50, and for the second it will be 100 × 0.1 or 10.

In this formula for exponential growth the growth rate also depends on N. When N is small the growth rate is slow and when N is large the growth rate is rapid. Thus if r is 0.5, and population size is 10, then dN/dt will be 5. If population size is 1000 then dN/dt will be 500.

The formula above gives growth rate in an exponentially growing population. If we wish to look instead at the population size at various times during exponential population growth, an equivalent expression is the integral equation

$$N_t = N_0 e^{rt}$$

Here N_t is number at time t; N_0 is number at time 0, the beginning of the period we are studying; e is the base of natural logarithms (about 2.718); r is biotic potential; and t is the time period being studied. The new small calculators with e^x keys make this formula easy to use directly. Formerly it was often easier to take the natural logarithm of each side of the equation and use the resulting formula:

$$\log_e N_t = \log_e N_0 + rt$$

Population growth rate in the logistic equation is given by

$$dN/dt = rN \left(\frac{K - N}{K} \right)$$

This is the same as the formula for exponential population growth except that the term $\left(\dfrac{K - N}{K} \right)$ has been added. K is known as the carrying capacity, and the whole expression $\left(\dfrac{K - N}{K} \right)$ is a measure of environmental resistance or the effect of crowding. Notice that when N is small $\left(\dfrac{K - N}{K} \right)$ is close to 1, so that biotic potential is nearly completely realized and population growth can be rapid. Suppose that $r = 1$ and $K = 50$. Then, when $N = 5$,

$$dN/dt = 1 \times 5 \times \frac{(50 - 5)}{50} \text{ , or } 5 \times 0.9$$

and dN/dt, then, is 4.5.

When N is large, growth is slow. If $N = 45$, then

$$dN/dt = 1 \times 5 \times \frac{(50 - 45)}{50} \text{ , or } 5 \times 0.1$$

dN/dt in this case is 0.5.

$\left(\dfrac{K - N}{K} \right)$ is a measure of environmental resistance but is an inverse measure; when environmental resistance is low, the numerical value of

$\left(\dfrac{K-N}{K}\right)$ approaches 1 and when environmental resistance is high the value of $\left(\dfrac{K-N}{K}\right)$ approaches 0.

We have just discussed growth rate during logistic growth (the bottom graph in Figure 3–7). To determine population size at any particular time (the top graph in Figure 3–7) we would use the equivalent integral equation

$$N_t = \frac{K}{1 + e^{a - rt}}$$

The terms are all the same as before except for a. Basically a specifies how close the population is to the carrying capacity when you start out. (More details on using the logistic equation, including how to calculate a, are given in Gause's *The Struggle for Existence.*)

POPULATION DENSITY AND POPULATION REGULATION

It is often convenient to discuss population size in terms of *density,* numbers per unit of space. Five hundred rabbits in a square mile is very different from 500 rabbits in a square block, and if we express the population size as rabbits per acre or rabbits per hectare we can make the comparison much more readily (500 rabbits per square mile would be about 1 per acre; 500 per square block would be about 250 per acre).

Many organisms show a similar density from one year to the next, measured from June to June or from October to October. By this we do not mean that their numbers are exactly the same every year; nature is more variable than that. The population does not, however, grow exponentially year after year, nor does it decline to extinction (Fig. 3–9). And for some kinds of organisms the numbers from one year to the next may be very close. Over a period of 18 years the number of red-eyed vireos breeding in a 65-acre beech–sugar maple forest near Cleveland, Ohio, studied by A. B. Williams, never went below 18 nor exceeded 36.

This type of stability leads us to believe that the size or density of some local populations is regulated by *density-dependent factors,* a term first used by the economic entomologist H. S. Smith. These are factors which vary in their effect according to the density of the population. Specifically, we are interested in factors that vary in such a way as to lower population size when the population begins to become too large or to allow population size to increase when the population begins to become very small (Fig. 3–10).

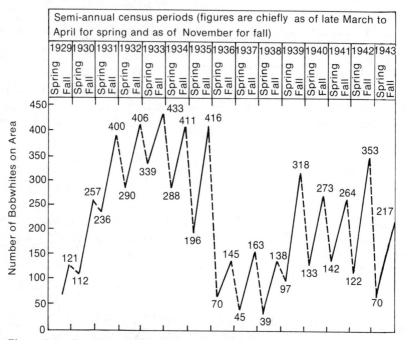

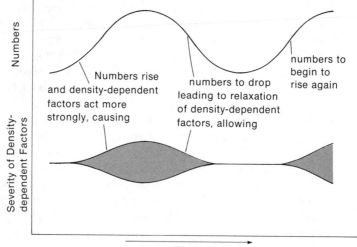

Figure 3–9. Population sizes of the bobwhite on an 1800-hectare (4500-acre) area in Wisconsin over a 14-year period (1929–1943). Frequent inspection of figures from real populations such as this one are desirable in population ecology to keep one's ideas grounded in reality. This graph shows that, despite our discussion of populations being regulated, the breeding (spring) population size in the bobwhite varied from below 50 to nearly 350 over five years. A second fact that is sometimes forgotten when looking at smooth theoretical curves is that populations living in strongly seasonal environments usually have a season of increase, when deaths are much overbalanced by births, and a season of decrease, when there are few or no births but deaths continue or increase. This is well illustrated here; the solid lines indicate spring to fall increases and the broken lines indicate fall to spring decreases. (From P. L. Errington, "Some contributions of a 15-year local study of the northern bobwhite to a knowledge of population phenomena," *Ecological Monographs*, 15:8, 1945.)

Figure 3–10. Basic features of population regulation by density-dependent factors.

An approximately analogous system would be the control of room temperature by an air conditioner. As the temperature increases, the thermostat turns on the air conditioner to cool the room. A little later, as the temperature drops, the air conditioner shuts off. When the room temperature rises again this series of events is repeated. This general type of control is called *negative feedback*. In the analogy room temperature is, of course, population size or density and the air conditioner is the density-dependent factor.

As density rises in a population, therefore, various factors come into play which raise the death rate, lower the birth rate, or raise the emigration rate. Table 3–4 illustrates this effect. When density falls, density-dependent factors allow the birth rate to rise, the death rate to fall, or the immigration rate to rise.

In talking of population regulation we are dealing with two closely related but slightly different questions. One is, what limits population size, or what sets the carrying capacity? The carrying capacity is approximately equivalent to the thermostat setting in our air conditioner analogy. The second question is, what are the actual factors that change in a density-dependent fashion (the factors of environmental resistance)?

Usually the carrying capacity seems to be determined by one or another of the following: climate, nutrition, or favorable space. Nutrition

Table 3–4 COMPARISON OF DENSITY-DEPENDENT AND DENSITY-INDEPENDENT MORTALITY ON POPULATION REGULATION

This table compares the effect on population size of a mortality factor that increases the proportion of the population killed as the population size (density) increases with a factor that kills a constant proportion no matter what the population size. In one sense, the second kind of factor is also density-dependent, since the number killed goes up with increased density.

Population Size at Beginning	Number Killed	Percentage Killed	Population Size at End
Density-Dependent Mortality			
100	0	0	100
150	50	33	100
200	100	50	100
250	150	60	100
300	200	67	100
Density-Proportional Mortality			
100	50	50	50
150	75	50	75
200	100	50	100
250	125	50	125
300	150	50	150

In the first case, the population size is regulated at 100 by the mortality factor killing an increased percentage with higher densities. The factor which kills a constant percentage does not have a regulatory effect. A second kind of density-independent factor, in which a constant number is killed no matter what the population density, could exist. It would not have a regulatory tendency either.

includes food (Table 3–5) and also various other kinds of nutrients, such as phosphate for lake algae.

Favorable space covers a multitude of virtues, such as nest sites, escape cover, and suitable soil texture for plants. If we find that by putting up more and more birdhouses we steadily increase the number of a bird species, we can conclude that nest sites had been the limiting factor for population size. If eventually, despite our putting up more nest boxes, the population size does not rise anymore we can conclude that some other factor is now limiting, and we would need to devise other experiments to find it.

The factors that act in a density-dependent way to regulate population size can be classified as *extrinsic* and *intrinsic*. Intrinsic factors are the population's own response to density; extrinsic factors involve interaction with the rest of the community. The main extrinsic factors are predation, parasitism, disease, and interspecific competition. It is hard to produce a completely satisfactory classification of intrinsic factors because they form a tightly interrelated group of processes; for purposes of discussion we will say they include intraspecific competition, immigration and emigration, and physiological and behavioral changes affecting reproduction and survival.

All the extrinsic factors can act in a density-dependent way and probably all of them, at some time for some populations, are important in regulation of population size. Clearly disease can spread more rapidly in a dense population because of the greater number of contacts between individuals. A dense population may be more vulnerable to predation because predators will be drawn to the area of concentrated food and because some members of the prey population will be living in poor cover. On the whole, extrinsic factors may be more a little more chancy

Table 3–5 THE INFLUENCE OF FOOD ON CARRYING CAPACITY*

The work of G. F. Gause on *Paramecium* is one of the classic studies of competition but we mention it here for another purpose. Gause cultured the two species separately in a water and salts solution and fed them daily with bacteria. The amount of food given was either one loopful or half a loop. The carrying capacity with half-loop and one-loop food supplies for two species are shown in the table.

Species	Food Supply	
	Half-loop	One-loop
Paramecium caudatum	64	137
P. aurelia	105	195

Thus, doubling the food supply while maintaining other conditions constant approximately doubled populations for both species.

*Data from Gause, G. F., *The Struggle for Existence*, Baltimore, Williams & Wilkins, 1934, p. 104.

than intrinsic factors. The predators that could be taking the surplus animals may be concentrated in some other area where the prey population has reached a still higher density; teleologically speaking, there may not be a predator around when you need one.

For many kinds of organisms intraspecific competition is one of the most generally important density-dependent factors. The word *competition* comes from roots meaning "to seek together" and is usually defined as a combined demand in excess of the immediate supply. Competition may be *intraspecific,* in which individuals within one species compete, or *interspecific,* in which individuals of two or more species compete. Interspecific competition, which may sometimes be important as an extrinsic density-dependent factor, has another important ecological role to be discussed later. For the moment we will concentrate on intraspecific competition.

Competition may be simply a matter of each individual taking as much as it can get of some resource, called *exploitation.* A thicket of young trees compete in this way; for a time there is enough sunlight or water for all of them to survive, although they do not grow as fast as they would if they were less crowded. Later some begin to lose out; they die, and the density drops.

In many kinds of animals, and in some plants and microorganisms, competition is more a matter of direct interaction than a matter of one individual using up a resource. This type of competition is called **interference.** Among animals, a prime example is **territoriality** (Fig. 3–11). An example in plants is the release of inhibitory chemicals, discussed in Chapter 4 in the section on *Chemical Ecology.*

In many different kinds of organisms, an individual defends an area, called a territory, against other members of the same species. Much of the singing of birds advertises the fact that the male is in possession of the piece of land. There are many different kinds of territories but if we take the simplest case, a territory on which a pair mates, nests, and finds food for themselves and their young, it is clear that for a particular area density of that species can be regulated fairly precisely. If there is room for about five pairs of scarlet tanagers in a 20-acre oak forest, there is likely to be about five pairs there, no matter how many are seeking to occupy the space. If numbers are high there will be some chasing, a lot of singing, and perhaps a little actual fighting early in the spring. Within a few days the situation will settle down, with about five pairs occupying territories. What of the other birds, the surplus? Several possible fates await them, most of which will reduce possibilities for survival and production of young as compared with the birds who won out. They may emigrate to less favorable habitat and breed, probably producing relatively few young because the habitat *is* unfavorable. They may not be able to establish territories at all but may simply form a floating, nonbreeding population.

Territoriality is social behavior. It involves communication and re-

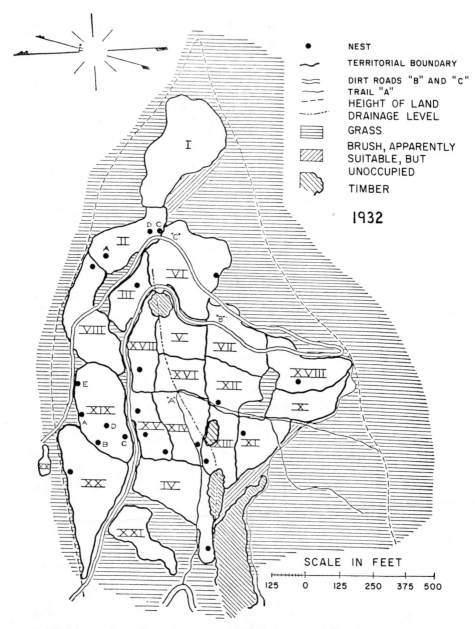

Figure 3–11. One of the first thorough studies of territoriality was by Mary Erickson on a small strange bird called the wren-tit, the most characteristic bird of chaparral. The map shows a chaparral-filled "gulch" near Berkeley, California, fully occupied by wren-tit territories an acre or two in size. (From M. M. Erickson, "Territory, annual cycle, and numbers in a population of wren-tits (*Chamaea fasciata*)," *University of California Publications in Zoology,* 42(5):255, 1938. Published 1938 by The Regents of the University of California; reprinted by permission of the University of California Press.)

Illustration continued on following page.

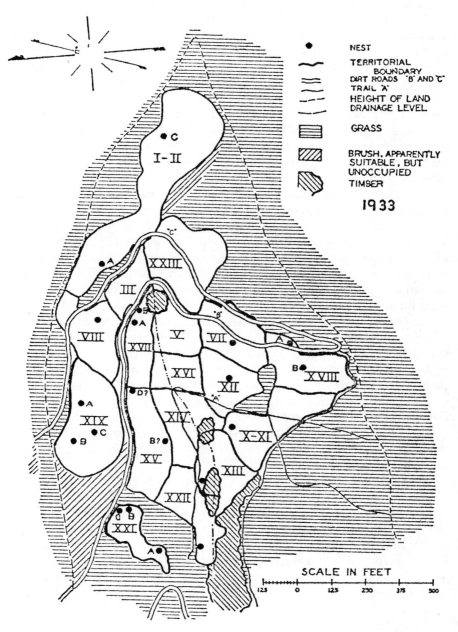

Figure 3–11 continued.

action among members of a population. The other intrinsic factors regulating population size are also socially based. Many organisms maintain no territories or perhaps only very small ones and consequently several individuals may have overlapping ranges; the number of individuals whose home ranges do overlap will tend to increase as density increases.

In some laboratory situations "social pressure" increases as density goes up and has some far-reaching physiological and behavioral effects. Just what social pressure consists of is not entirely clear, but it may include factors as simple as the frequent disturbance that occurs when things are crowded or as complicated as competition for rank in the social order (that is, which animals are to be dominant). Physiological pathways may involve endocrine glands such as the anterior pituitary and the adrenal cortex, but we will leave these details to physiologists and concentrate on effects which could lower population size. A partial list of these includes lowered resistance to disease, failure of females to build proper nests or to nurse young, and even failure of males to produce sperm or females to come into heat. It is certain that these effects befall populations when they get high enough, as in enclosed laboratory situations. What is not clear is how far along this road natural populations go. Fairly often social pressure probably leads to emigration before some of the more dramatic results occur.

In speaking of population regulation and limitation we must consider local populations, such as we have just been discussing, and regional or whole species populations separately. Population size of a whole species or the regional population size is probably limited but not regulated. Limitation of numbers of a species is, in the broadest sense, based on the smallest amount of favorable habitat available during the year. There seems to be no way for the evolution of *regulatory* mechanisms for a whole species to occur. Instead the species which do not happen to have the population traits that allow them to avoid extinction are, by definition, no longer with us.

We can readily see how factors regulating the size of local populations might evolve. It is evolutionarily advantageous for animals to be able to recognize crowding and to avoid it, if possible. For example, the regulation of local populations of birds occurs by individuals giving signals which allow one another to assess the degree of crowding. As areas become crowded newly arrived individuals will tend to go elsewhere, to areas that may be physically less suitable but that may be more favorable for them because they are uncrowded.

GROWTH AND REGULATION OF HUMAN POPULATIONS

The *crude birth rate* and *crude death rate* are the simplest figures to use for studying human population change. Crude birth rate is the number of births per thousand persons in the population. If there were 250 children born to a population of 10,000 during a year, then the yearly crude birth rate would be 25. Crude death rate is calculated in the same way; if 150 persons died in the same year then the crude death rate would be 15 per year.

Such a population is growing because birth rate exceeds death rate; determining how fast it is growing is simple enough. There were 10 more births than deaths per thousand persons (a crude birth rate of 25 per thousand minus a crude death rate of 15 per thousand). Expressed as a percentage (per hundred rather than per thousand) this is an increase rate of 1% per year. There were 10,000 in the population last year; there were 150 deaths and 250 births, for a net increase of 100. This 100 is the 1% increase over the original 10,000.

For a stable population, of course, the rate of population increase would be zero, since birth rate would be equal to death rate. If a population grows at a fairly constant rate, whether it is 1%, 5%, or ½ of 1%, population size will increase exponentially. This is exactly the same situation that we discussed earlier in more general terms (in the section on *Exponential Population Growth*). The only difference is that the rate of increase may be slower or faster than would be predicted on the basis of the biotic potential of the organism.

One way to understand the consequences of exponential growth resulting from even low percentage rates of increase is to consider the relationship between the yearly rate of increase and the number of years it will take the population to double in size, assuming that the yearly rate continues. Some of these figures are given below:

If the yearly percentage increase is	The years required for the population to double will be
0.5	139
1.0	69
1.5	46
2.0	35
2.5	28
3.0	23

For the population that we have talked about, with a growth rate of 1% per year, the population would double in size in less than 70 years. (As an approximation, good enough for everyday purposes, you can calculate doubling time by dividing the percentage increase into 70. This works as long as the percentage is below about 15.)

Throughout much of human history population size stayed steady or increased slowly as new lands were discovered and colonized. Fairly high birth rates were approximately balanced by fairly high death rates. Ten thousand years ago, as the last Ice Age ended, it has been estimated that there were no more than 5 million humans. With the rise in agriculture and the shift away from the hunter-gatherer way of life a cycle of population growth was begun. World population may have first reached 100 million shortly before the time of Christ, or somewhere between 1 and 2 million years after man first appeared on earth. The world population had reached 500 million about 1650 A.D., when Rembrandt was painting, Leeuwenhoek was inventing the microscope, and Louis XIV was doing whatever kings of France did in those days.

In the following years world population growth was accelerated by events associated with industrialization, urbanization, and, particularly, the exploitation of sparsely occupied lands in North and South America and Australia. By 1850, about the time Henry Wadsworth Longfellow was writing *The Song of Hiawatha*, Abraham Lincoln was practicing law in Springfield, and Charles Darwin was working on barnacles, the world population had reached 1 billion. The next billion took 80 years, to 1930, the third billion took 30 years, to 1960, the fourth billion about 15 years, and we are already a third of the way to the fifth billion.

In many hunter-gatherer societies a reasonably close regulation of local population size was achieved, with both physiological and behavioral factors probably involved. Prolonged nursing of young probably had some effect in suppressing ovulation, tending to keep women infertile for 3 or 4 years after a birth. Postpartum sexual abstinence, induced abortion, geronticide, and infanticide were all practiced. Infanticide, apparently, was particularly widespread. Expulsion and immigration also tended to maintain local population size below the carrying capacity of the band's territory.

In regions suitable for the growth of crops agriculture raised the carrying capacity of the land. Once a sedentary village life was adopted transporting infants was not a problem, so there was no longer that penalty for shortening the intervals between births. In fact, most penalties for overreproduction were reduced for the individual parents because child care was spread over extended families. Along with the fact that large families are often an economic asset in agricultural societies, these features seem to explain the jump in population growth after the agricultural revolution. It would be a mistake, however, to think that no preindustrial agricultural societies regulated their population size. Some did, by delaying the age of marriages and, especially, by restricting the inheritance of land to the eldest son.

In the eighteenth and nineteenth centuries, as industrialization occurred in western Europe, there occurred a series of events that sociologists have called the **demographic transition.** What happened was approximately this: There was a decline in death rates, initially through better sanitation and attention to public health. (One of the first public health laws in England made it a capital offense for anyone restricted to his house with the plague to break quarantine.) Later, in the nineteenth century, improvements in medical practice also became important. With the decline in death rates, there was a burst of population growth. England's population increased from not much more than 10 million in 1815 to about 29 million in 1890 while, at the same time, about 11 million emigrants left for the U.S. and elsewhere.

Birth rates also dropped but later and more slowly, and rarely low enough to balance death rates. A typical pattern was for the death rate to drop from 40 to 50 per thousand to about 10 per thousand while the birth rate dropped only to about 15 per thousand. The University of Oregon anthropologist D. E. Dumond has suggested that birth rates dropped because, in an industrial society, many of the responsibilities of child care were returned to the nuclear family. The parents again bore the cost of rearing children at the same time that the rise of compulsory schooling decreased their economic value (as wage-earners).

This pattern of the demographic transition was repeated in many countries as they industrialized. What of the unindustrialized nations? Until shortly before World War II they continued the primitive pattern of high birth and death rates. Sanitation and medicine were exported to these countries beginning about that time and at an accelerated rate after World War II, with a consequent decline of death rates. Birth rates have not followed suit, so that at present population growth is extremely rapid in the "developing" countries. From 1850 to 1920 the increase in the world's population was about equally divided between industrialized and developing parts of the world. Now, however, about 80% of the world's population increase comes from the nonindustrialized nations.

Following are some figures for percentage population growth for various countries from the 1977 World Population Estimates as prepared by the Environmental Fund:

Country	Percentage growth per year
Brazil	2.9
Mexico	3.5
India	2.3
Uganda	3.3
Colombia	3.2
Japan	1.1
United Kingdom	0.0
Poland	1.1

It is worth looking at recent figures for the U.S. separately. In 1960 the percentage growth rate was 1.6; in 1965, it was 1.2. By 1970 it had dropped to 1.1 and for 1977 it was about 0.5 based on a crude birth rate of 14 and a crude death rate of 9. The number of children per woman had dropped to a level that would allow population stabilization at some point 40 or 50 years in the future. (Remember that, owing to the large number of young persons in the population, birth rate will remain higher than death rate and the population will continue to grow for some years, even though family size is at replacement level.)

However, there is another complication here. Immigration is as much a reality in human populations as any others. The U.S. Immigration and Naturalization Service estimated that there are now more than one million illegal immigrants every year (and 400,000 legal ones). Growth from births is about 1,300,000 or, in other words, immigration in the U.S. is now contributing as much or more to population growth as reproduction. The actual population growth rate for the U.S., accordingly, may be as much as 1.5% rather than only about 0.5%.

Any precise prediction that we make about the future population of the U.S. or the world is likely to be wrong. Such predictions are merely statements of what will happen if growth rates stay the same, but history has shown that growth rates are changeable (Fig. 3–12). In 1935 the U.S. growth rate was about 0.8, nearly as low as it is now. In 1956, it was 1.8, three times what it is now. It may increase again, starting the country on a new wave of population growth, or it may drop still lower and halt population growth by the year 2000.

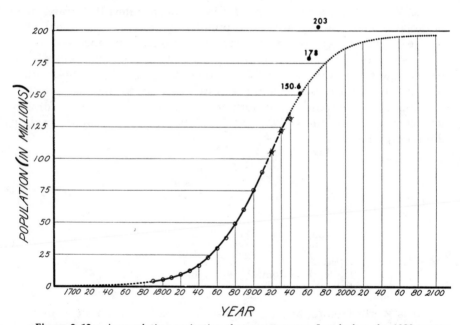

Figure 3–12. A population projection that went wrong. Just before the 1920 census Pearl and Reed found that population growth in the U.S. from 1790 to 1910 seemed to have taken the form of a logistic curve which would tend to level off below 198 million somewhere after the year 2100. The 1920 census figures fit fairly well, as did the 1930 figures. The 1940 figures, however, definitely did not fit and Pearl and his coworkers wondered if the curve needed revision. They revised the curve downward, predicting an upper limit of around 184 million. We know now, of course, that the slowing in growth resulting from the drop in birth rate during the 1930's depression was temporary. As shown by the points added to the curve, the 1950 data was close to their projections but by 1970 the U.S. population was already far greater than their projected limit of around 198 million. None of this is a reflection on Pearl or his colleagues, who clearly stated the limitations of any population projection. (From Raymond Pearl, L. J. Reed, and J. F. Kish, *Science*, 92:487, 1940.)

As for the world population (we need not worry about immigration and emigration here), it has been growing at about 2% per year, which would give a doubling from the present 4 billion to 8 billion in 35 years. We may hope that this prediction will be wrong, on the high side, but one imprecise prediction that we can make with confidence is that the world population is going to get larger before it gets smaller.

ORGANIZATION IN POPULATIONS

SPACING

Spacing (also called *pattern*) refers to the positions of members of a population relative to their neighbors. There are three basic kinds of spacing, *random, clumped,* or *even* (Fig. 3–13).

To illustrate the three types, visualize a long lunch counter with 50 stools with 10 persons on them. If there is no pattern to their spacing then we say that the distribution is random. Specifically, it is random if every stool has an equal and independent likelihood of being occupied. This may well be true if each person came in separately and sat down where it was convenient to do so. But if the 10 persons came in as two groups, perhaps two families, then each group would probably sit on stools next to one another. This distribution, in which individuals tend to occur in bunches or clumps with empty spots between, is a clumped (also called aggregated) distribution. A flock of birds or a bed of flowers are other examples of clumped distributions.

Suppose now that we start out with 10 persons in a random distribution and 15 more people come in until there are 25 persons on 50 stools. Although there are friendly types who will sit down next to you and start a conversation, most people tend to leave some space between them and their neighbors. Consequently, the pattern at the lunch counter would probably look like this: person, stool, person, stool, person, etc. This is an even distribution, in which individuals are spaced more regularly than in a random distribution. Other examples of even distributions are the trees in an orchard or the honeybee larvae in a comb. In the honeycomb we see the most extreme case of even spacing, in which each individual is equidistant from six others.

It is reasonable that individuals should be randomly spaced unless

Figure 3–13. The three general types of spacing.

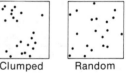

Even   Clumped   Random

something is happening to warp the spacing toward being uniform or aggregated. Actually random spacing is not very common in nature, nor is even spacing. By far the most common pattern is clumping.

Several factors may operate in different populations to produce clumping. First, variation in the spatial distribution of environmental factors, such as moisture, food, or cover, is often patchy. If a wildflower grows only in the moister parts of the forest it is apt to occur in clumps in the low spots. Second, dispersal patterns could cause clumping. For example, many plants reproduce vegetatively by spreading underground stems or by some similar method. We may find well-defined, nearly circular clumps of May apple, for example, the result of vegetative spread from the single individual that started years before from seed. Third, for many animals, social behavior is responsible—that is, the animals associate with other animals like themselves.

Factors tending to produce an even distribution are mainly competitive interactions between individuals. The territoriality of birds causes them to be evenly spaced (this is, of course, no less a social response than flocking). The trees making up a forest canopy may be on the even side of randomness as the result of competition for sunlight (Table 3–6).

One final point about spacing is that a population may show one pattern at one scale and another at a smaller scale. Herds of wildebeest

Table 3–6 DENSITY AND SPACING IN A SAND-PINE
SCRUB COMMUNITY*

The sand pine scrub is a unique chaparral-like community in Florida that has been studied by A. M. Laessle. It is maintained by fire and the dominant species, sand pine, has serotinous cones like those of jack pine and lodgepole pine. After a fire, a dense even-aged stand of small trees grows up. Their spacing is random or slightly clumped; with time, density declines and the spacing becomes even.

Time since Burnt (years)	Density (per m²)	Spacing
12	2.6	Random
20	0.5	Random
25	0.2	Even
51	0.08	Even
66	0.04	Even

The tendency toward even spacing is caused mainly by the death of trees, from competition with their neighbors. Deaths could keep the spacing random, if they occurred randomly, or they could produce a clumped distribution if the deaths occurred in patches. In fact, the deaths produce an even distribution, showing that crowded trees are more likely to die than uncrowded ones. This is shown below, where data from the 55-year-old stand compares the spacing of all trunks (alive plus still-standing dead) with the live trunks.

Trunks Measured	Density (per m²)	Spacing
Alive plus dead	0.16	Random
Alive only	0.08	Even

*Data from Laessle, A. M., "Spacing and Competition in Natural Stands of Sand Pine," *Ecology* 46:65-72,1965.

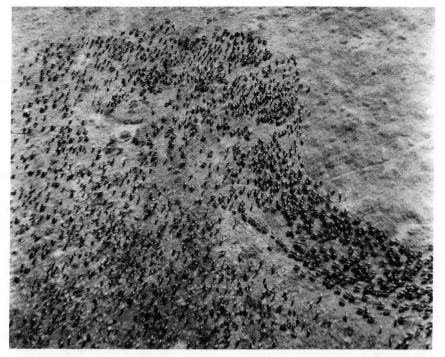

Figure 3–14. These migratory wildebeests occur in aggregations on the Serengeti plains. Within the herds, however, the animals tend to be evenly spaced, maintaining an individual distance from one another. (From *Science*, 191(4222): cover, 1976. Photograph by S. J. McNaughton.)

in African grasslands are clearly aggregated, but if our scale is not one of several square miles but a few hundred square feet, the picture changes (Fig. 3–14). If we look just at spacing within the herd we see even spacing because of the tendency of the animals to maintain a certain minimum *individual distance* between themselves and their neighbors. This is much the same phenomenon that we saw at the lunch counter. If we see a flock of swallows on a telephone wire they are not wing to wing; rather, there is a definite distance separating each bird. Infringement of this circle of personal space by another bird will cause the first bird to shift away or, in the appropriate circumstances, to fight. The circumstances in which close contact between two individuals is tolerated or welcomed vary between species, but two fairly widespread cases are mating and care of dependent young.

MATING SYSTEMS

There are three basic mating systems in a population, *monogamy, polygamy,* and *promiscuity.* In monogamy a persistent pair bond is formed between one male and one female. Polygamous unions involve a bond between a single member of one sex and several of the other sex. The most common type of polygamy, at least in birds and mammals, is *polygyny*, in which one male has a harem of several females. *Polyandry*, in which one female has several male consorts, is rather rare. In promiscuity the individuals meet for copulation but no continuing relationship is established.

Monogamy lasting through a brood or a breeding season is the usual situation with birds. Longer unions lasting several years or up to life are known for many hawks, swans, geese, crows, and chickadees. Among mammals, carnivores are monogamous. Polygynous animals include a good many birds such as the red-winged blackbird, most pheasants, and the ostrich; polygynous mammals include many of the primates, such as baboons (however, gibbons are monogamous and chimpanzees promiscuous) and some of the large grazing animals such as deer. Polyandry is known for a few birds and is usually accompanied by a reversal of sexual roles, the male incubating the eggs and caring for the young. Promiscuity is the usual condition among smaller mammals and also occurs in some birds, notably *lek* species. These are birds in which the males gather in an area (referred to as a lek or an *arena*) and go through various displays. Females visit the lek for copulation and go elsewhere in the vicinity for nesting. Grouse, such as prairie chickens and sharp-tailed grouse, are examples.

There are good ecological and evolutionary reasons for the existence of various mating systems. Female mammals, unlike female birds, are not tied to a nest and eggs that require incubating and defense from predators. Furthermore, they produce milk as a food supply for the young once they are born. For these reasons, then, the continued presence of a male is not a necessity in mammals. In fact, by eating food that the female could turn into milk and by attracting the attention of predators, the male may be a liability after impregnation has occurred. Hence, promiscuity has evolved as a mating system among many mammals. For most birds care of the nest and young by both parents is advantageous and most species are monogamous. The monogamous mammals are the hunting mammals, in which the male is useful for obtaining meat for the female and young.

This discussion applies, of course, primarily to vertebrates. Many of the sexual and reproductive arrangements in invertebrates are not readily classified by vertebrate standards.

AGE STRUCTURE

The age structure of a population refers to the proportions of individuals of various ages. Age may be expressed in days, months, or years, or the categories prereproductive, reproductive, and postreproductive

Ages

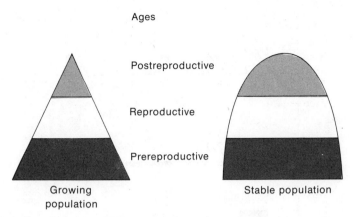

Postreproductive

Reproductive

Prereproductive

Growing
population

Stable population

Figure 3–15. Diagrammatic representations of age structure in a growing population and a stable population.

may be used. An exponentially growing population takes on a characteristic age structure, called the *stable age distribution.* A population which is not changing in size has a different age structure, in which there are more older and fewer younger individuals than in the stable age distribution; called the *stationary age distribution,* it is calculated from the l_x column in the life table. A graph of age structure in an exponentially growing population is triangular whereas in a stationary population, it is somewhat bell-shaped (Fig. 3–15). Any real population usually has an age structure different from either of these because of various events in its recent past (Fig. 3–16 and Table 3–7).

Table 3–7   AGE DISTRIBUTION FOR THE BLAZING STAR*

Age structure in plant populations has not been studied for many species. Below is the age distribution for a herb of dry prairies, a blazing star. One age distribution is for the species on undisturbed sites; the other is for sites buried under sand. Age was approximated by digging up plants and counting annual rings in the underground stems.

Age (years)	Undisturbed Sites	Sites Buried under Sand
1–4	5%	65%
5–8	8	11
9–12	22	10
13–16	24	11
17–20	24	2
21–24	8	<1
25–28	5	0
29–34	3	0

The conclusion is that the persistence of blazing star is based on occasional disturbance, such as sand deposition. Without this few new individuals are successfully established and the old ones eventually die.

*Data from Kerster, H. W. "Population age structure in the prairie herb, *Liatris aspera,*" *BioScience,* 18:430–432, 1968.

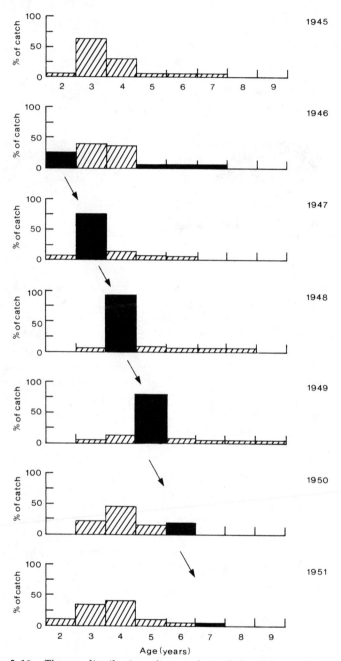

Figure 3–16. The age distribution of any real population has peculiarities based on events in the recent history of the population. The graphs show that the 1944 year class (*black bar*) of whitefish in Lake Erie was unusually large, and this event warped the age distribution for several years. The figures are from fish taken in commercial fishing. The age distribution of humans in the U.S. would show peculiarities based on the low birth rates of the depression years, the high birth rates of the "baby boom" days following World War II, and the low birth rates that have marked the past two or three years. (From C. J. Krebs, *Ecology*, New York, Harper & Row, 1972.)

Age structure is important in population growth. We can visualize two sets of sailors who have mutinied, cast their captain adrift, and set sail with a bevy of South Seas maidens for an uninhabited island to found a new colony. In one case imagine the sailors and maidens as all being about 19 years old and in the other about 55 years old. The potentiality for exponential growth is high in the first case (higher, in fact, than would be predicted from the biotic potential which assumes the stable age distribution) and very low in the second. The presence of a high proportion of persons who have just entered or are just entering reproductive age is responsible for the fact that, even if the average number of children per family moved to the replacement figure of just over 2 and stayed there, the population of the U.S. would continue to grow until well into the next century, when it would stabilize (assuming no immigration) at about 275 million.

The average age of an individual in the U.S. today is about 29. If the population stabilized, with death rates remaining about the same as now, the average age would go up to about 38 halfway through the twenty-first century. This is an inevitable consequence of population stabilization—inevitable unless stabilization is achieved by raising death rates rather than by lowering birth rates, in which case a young population could be maintained. The average age has moved up fairly steadily through the years; in 1800 it was about 16, in 1900 about 23. With stabilization, the proportions of various ages will shift. There will be slightly more persons over 50 than under 20, whereas now the youngsters outnumber the oldsters 3 to 2.

The many possible economic and sociological consequences of this change in age structure are fascinating but too far afield. It is evident that a greater proportion of our resources will have to be used for helping the elderly and a smaller proportion used for education and other child care services. However, the total economic burden should be no greater because the "dependency ratio" (the number of nonworkers as compared with the number of workers) may be slightly less. In Table 3–8, for example, it can be seen that 42% of the present population in the U.S. is either under the age of 17 or over 65. With a stationary population in 2050

Table 3–8   APPROXIMATE AGE DISTRIBUTION IN THE UNITED STATES AT THE PRESENT TIME AND IN 2050*

Age Class (years)	Today	2050
0–17	31.8	24.0
18–44	37.5	35.3
45–64	20.4	23.7
65 and over	10.3	17.0

*Assuming a stationary population size of about 275,000,000. From *Statistical Abstract of the U.S.*, 97th ed., Bureau of the Census, 1976, p. 6–7.

A.D., the total of these two groups will constitute only 41% of the population.

EVOLUTION

NATURAL SELECTION

Organisms evolve. An ecological problem for an organism today may not be a problem a few generations from now because the organism has the potential to change, through natural selection, so that it becomes adapted to the situation. As currently viewed, natural selection is only very slightly different from the way Charles Darwin described it in 1859. Individuals vary in a population. Those whose variations suit them best to the existing environment tend to leave behind more descendants in the next generation, because they survive longer and reproduce more. Consequently, any of the favorable variations that are hereditary will tend to accumulate in succeeding generations; in other words, a larger and larger percentage of individuals of the population will possess these traits. The percentage of individuals possessing the unfavorable traits will decline because in each generation they die early or have few offspring. Evolution, then, in its simplest sense, is a change in the proportions of individuals in a population possessing some given hereditary factor, or gene. Note that evolution is a process of *populations*. Individuals do not evolve; they survive a longer or shorter period and leave a larger or smaller number of descendants according to the relative favorability of their genetic makeup.

The ultimate source of the genetic variability important in evolutionary processes is presumably mutation. However, at any one time, natural selection works mainly on the variability already present in the population. It has been discovered in the last few years that most populations have a lot more variability than previously realized.

Most mutations that have been studied are disadvantageous in some way. At first this may seem puzzling. How can mutation be important in adaptation if mutations are harmful? The explanation seems to be this: Most species have been around for hundreds or thousands of generations, so that most mutations that are likely to occur *have* occurred. The favorable ones have already been incorporated as one of the normal conditions of the organism; the harmful ones tend to be continually eliminated. It is these harmful mutations, as they crop up again, that we detect as new.

This situation prevails as long as the environment of the organism does not change materially. Various insect populations possess mutations that, as one of their effects, make the insects resistant to DDT. In a pre-DDT environment this effect was not of any use to the insect and was usually coupled with some other effect that lowered the fitness of the insects possessing the gene. In the DDT era, of course, these genes became extremely advantageous. A good spraying of DDT would all but wipe out most of the insect population, leaving the environment uncrowded for the few resistant individuals. These would prosper to such a degree that only a few years would be enough for a population to evolve from a situation in which nearly every individual was killed by DDT to one in which practically none was (Fig. 3–17).

Some have suggested human evolution as a solution to our environmental problems. If some pesticide is injurious to us today, we will evolve immunity to it. If the genes for such immunity occur in the human gene pool or if mutation should happen to produce them, then this is a possibility. Few biologists would regard it as a feasible or humane solution, though, because they know that to produce a largely immune population in a few generations would require the death (or at least the nonreproduction) of nearly every person who was not immune; anything less drastic would take many generations.

R AND K SELECTION

The traditional view of natural selection is one in which biotic potential or the intrinsic rate of natural increase (r in the logistic equation)

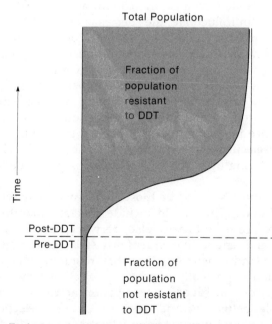

Figure 3–17. Evolution of resistance to DDT in an insect population. In a pre-DDT environment some small fraction of the population contains genes giving some measure of resistance to the insecticide. When applications of DDT kill most of the nonresistant individuals, the environment becomes uncrowded for the resistant ones. These reproduce heavily and survive well so that, very rapidly, the population changes to one in which a large proportion possesses genes for DDT resistance.

would be maximized. In other words, if a population has two genotypes differing in reproductive rate, then the one which reproduces faster will increase in numbers relative to the slower one. It must generally be true that genotypes favoring fewer eggs, later puberty, and the like lose out in natural selection. However, there is indirect evidence that evolution sometimes produces lower reproductive rates. For example, certain very large flying birds such as the California condor and various seabirds lay only one egg per year (or sometimes only one every other year) and require several years to become sexually mature. Unless we believe that the earliest birds had a single-egg clutch and deferred maturity, and that the condors almost alone among birds retained the combination, then we must conclude that they evolved from birds having a higher reproductive rate.

Robert MacArthur and E. O. Wilson have suggested that natural selection may take different courses in species that generally live under uncrowded conditions versus those that generally live at about carrying capacity (that is, at K in the logistic equation). They refer to selection under uncrowded conditions as *r selection* and selection under carrying capacity as *K selection.*

r-selected species would be those that live in short-lived or unstable ecosystems such as early successional stages, river sandbars, and forest openings. Here, either because the community is so short-lived or because conditions in it fluctuate, the species is rarely at K. Instead, it often lives under uncrowded conditions; competition, especially intraspecific competition, is rarely important. Some traits favored by natural selection under r selection are, first, several features favoring a high intrinsic rate of natural increase (r), including a large number of young, eggs, or seeds and reproduction at an early age. Small body size is often favored, with energy diverted from body tissue to reproduction. An *annual* life history—reproducing once in the first year of life and then dying—is sometimes the result of r selection. Also important for r-selected species is high dispersal ability, giving them the ability to colonize new favorable areas as they arise.

K-selected species would be those that generally live under carrying capacity conditions. This would include species inhabiting stable sites such as climax forests and caves; also, very large species that range over large areas of land tend to live under fairly stable conditions even though portions of the landscape over which they roam may fluctuate. Intraspecific competition is generally important. Selection may favor a lower r, with the energy or time saved being used for increasing competitive ability (for example, better territorial defense). Under K selection larger body size, a longer life, and increased specialization (in feeding, for example) seem to evolve.

One very good illustration of r-selected versus K-selected species is seen in a comparison of pioneer and climax plant species. In the earliest stage of plant succession, which lasts only about a year in the midwestern U.S., plants are widely spaced, suggesting that the area is not crowded. They are all herbs and mostly annuals which produce a large number of seeds, and many of the plants are wind- or animal-dispersed. (Several of the most important species, however, do not show any obvious means of rapid, distant dispersal. At first this may seem puzzling but further study shows that these species have an alternative way of attaining the same goal. Their seeds are very long-lived: annual ragweed seeds may live for 40 years, evening primrose seeds for 80, and moth mullein for 90. In the course of time, with an occasional ragweed plant growing here or there and producing seeds, seeds become very widespread in the soil. When an area is disturbed—a forest cut over, for example—ragweed does not need to disperse to the area; it is already there in the form of seeds which now find favorable conditions for germination.)

The dominant plants of the climax forest are not herbs but trees. They are large, live for tens or hundreds of years, and reproduce repeatedly. The herbs that are a part of the climax forest are also long-lived. In the climax forests of the eastern U.S. you can count the number of annual species on the fingers of one hand. Most of the climax forest herbs produce relatively few seeds but they reproduce readily by rhizomes and other vegetative means, so that when they reach a spot they are able to spread over and dominate it.

The concept of r and K selection is most useful when comparing related organisms. The insects of climax forest may be K-selected as compared with insects of crop fields. However, most of the insects of the forest may be r-selected if compared with the birds of the forest. And, of course, no species is totally r-selected or totally K-selected; what is found is a mixture of traits suited to the particular habitat and the particular position a species occupies in that habitat.

Possibly the concept of r and K selection seems remote from any practical use of ecology, but this is far from the truth. Consider the question of endangered or threatened species. Suppose that something (such as DDT) makes some proportion of the population sterile or, at least, prevents them from reproducing successfully. The results are very different in r-selected versus K-selected species, as was shown by the avian ecologist Howard Young. Table 3–9 compares an r-selected species (breeds at age 1, has a clutch of 5.6 eggs and a mean annual mortality rate of 73.8%) with a K-selected species (breeds at age 6, 1.3 eggs, 50% mortality first year, 12.5% each succeeding year). The particular species used as models are the American robin and the bald eagle. If 90% of the population fails to reproduce, population size in the two species decreases, as shown in Table 3–9.

For the r-selected species the decrease in numbers will be detected quickly, almost certainly by the second year, and the potential exists for doing something about it. In the K-selected species, however, almost no

Table 3–9 POPULATION DECREASE IN R- AND K-SELECTED SPECIES*

Year	r-Selected Species	K-Selected species
0	1,000,000	1,000,000
1	270,526	801,984
2	73,185	702,976
3	19,799	616,345
4	6,669	540,544

*From Young, Howard, "A consideration of insecticide effects on hypothetical avian populations," *Ecology*, 49:993, 1968. Copyright 1968 by the Ecological Society of America.

successful reproduction has gone on for four years and the population is not yet halved. Consequently K-selected species require much closer monitoring of their numbers than do r-selected species if we are to detect problems early enough to deal with them. Otherwise the K-selected population may become so top-heavy with old individuals that prospects for recovery are poor before we even realize a decline in population size is occurring.

On a more general level, it is of course primarily r-selected species that make suitable game animals. K-selected species rarely have the biotic potential to replace animals lost to hunting. Possibly the best example is the passenger pigeon, a forest species which laid one one-egg clutch per year. It was still fantastically abundant in the early 1800's, but by 1900 it was extinct in the wild.

SOCIOBIOLOGY

A *society* is a group of cooperating individuals of the same species. Many degrees of social behavior exist, from schooling in fish to the complex interactions within societies of ants, termites, chimpanzees, and man. The study of the biological basis for social behavior, whether of animals or man, has recently been called *sociobiology* by E. O. Wilson.

The origin and evolution of many features of the truly social, or *eusocial*, insects have long been puzzling. In these insects, which include all ants and termites and certain bees and wasps, there is a division of labor between reproductive castes which produce eggs and sterile castes which provide food, care for the developing larvae, and defend the colony. The evolution of sterility is difficult to explain by the usual view of natural selection, which is based on the reproductive contribution of individuals to succeeding generations. Sterile individuals obviously make no direct contribution to succeeding generations.

The explanation may be that a type of *group selection* is operating. Families or colonies of related individuals are more successful (we may suppose) if they contain sterile workers; consequently, the genes favoring the production of sterile workers by the queen will increase in the population through the founding of new colonies by new queens produced in those successful colonies. Because it depends on some individuals sacrificing fitness (in the genetic sense of contributing to the next generation) in favor of relatives, this kind of selection has been called *kin selection*.

Taking care of one's sisters (as social insect workers do) can be thought of as a form of *altruism*. Sharing food (humans and wild dogs) and defending one's country or colony (humans and soldier ants) are other examples. In altruistic behavior the individual potentially lowers his or her fitness while potentially raising that of the individual helped. Kin selection is one avenue for the evolution of such behavior. If we are altruistic to our relatives, the genes we share may increase (including the ones for altruism), even though one of our altruistic acts may shorten our own life.

Of course, closer examination may show that some selfless acts are selfish instead and thus require no special mechanism to explain their evolution. Many birds, when they sight a flying hawk, give a warning call (of an almost ventriloquial quality) that causes flockmates to seek cover. This behavior seems well-suited for a kin selection explanation: the calling bird may slightly increase his chances of being caught but greatly increases the chances of flockmates (mostly relatives) escaping. However, it may be that the warning enhances the warner's fitness directly and the help to others is merely a bonus. Suppose, for example, that some or all of the following are true: the hawk is unlikely to stop unless vulnerable prey are visible; if it does stop, it is unlikely to remain long unless it is successful or nearly so; if it does catch one animal, it is likely to stay in the vicinity for a long period of time, picking off the survivors. If these are true, the calling individual may be contributing to his own survival directly by getting his flockmates out of sight.

Part of Wilson's book *Sociobiology* deals "in the free spirit of natural history, as though we were zoologists from another planet completing a catalog of social species" with human social behavior and its evolution. To many scientists an understanding of the hereditary basis of human social behavior seems to have practical importance. It might indicate what human societies safely could and should not do in the future, under pressure for social change coming from increasing population size and technological innovations. Opposition to this approach has come, expectably, from some sociologists and anthropologists who argue that human culture cannot be studied as biology. Less expectably, perhaps, opposition has come from a few scientists who argue that any hereditary basis of human behavior should not be studied because such study might be used as justification for repressive political policies, such as racism and sexism.*

---

*This latter viewpoint is set forth in "Sociobiology—another biological determinism," *BioScience*, 26:182, 184–186, 1976.

PREDATOR-PREY RELATIONS

FEEDING AND PREDATION

The use of one organism as food by another is a complex subject. First of all, the activity is the basis of the transfer of energy through the community (see Chap. 4). Secondly, it is a subject that must consider the diverse kinds of hunting and foraging techniques used by animals to find and catch food, and the diverse kinds of behavior and structure by which plants and animals avoid being eaten. Third is the numerical effect, the influence on population size of both the eater and the eaten as a result of the interaction. Fourth is the evolutionary effect, the selective effect that predation has on the prey and that food shortage has on the predator.

There is no entirely satisfactory classification of feeding types of animals. Ways of feeding are exceedingly diverse ranging from clams that filter water for microscopic organisms and bits of organic matter to wolves that catch and eat moose, from bison that crop off grass to plant lice that suck sap out of leaves, from bird lice that eat feathers to blood flukes, liver flukes, or intestinal flukes that soak up nutrients from whatever kind of body fluid they live in. There seems to be a fairly basic subdivision between the organisms that routinely kill another individual organism in their feeding and those that get their food without necessarily killing the individual supplying it. The first category includes predatory animals such as insect-eating birds and rat-eating snakes, but also seed-eaters such as many kinds of birds, small mammals, and insects. The second category includes grazing and browsing plant-eaters and external and internal parasites.

If we simply take a particular animal as it is, with sharp teeth or blunt, long legs or short, a digestive system built for grass or one for flesh, then its diet probably depends mainly on two factors; the availability of different food items and the animal's food preferences. Of course, in starting at the present and ignoring how the horse came to be fitted for eating grass or the Everglade kite for eating a single kind of snail we are ignoring a great deal of ecology. The relationship between the eater and the eaten is a strong evolutionary force for both, and we will return to the topic a little later.

In the summer red foxes may eat meadow mice; in the winter when snow is deep they may eat rotting apples, since the apples are what is available to them. Availability depends on abundance but not strictly on abundance. Meadow mice may not be much less abundant in winter than in summer but they are much less available, much less easily found, when the snow is deep. Food items may be unavailable for other reasons; they may be camouflaged or have some other kind of protective coloration or shape, or they may have protective devices such as thorns, spines, or prickles. It is no accident that in an overgrazed pasture everything may be cropped off low except the thistles. The protective devices may consist of unpleasant or poisonous chemicals; most people know that toads and monarch butterflies are poisonous, but there are a great many other kinds of animals and plants that produce poisonous or repellent substances. Millipedes spray cyanide; the bombardier beetle through a most remarkable system shoots out noxious, boiling hot chemicals. Nicotine, caffeine, pyrethrin, and rotenone are examples of chemicals that seem to protect the plants producing them from grazers. Organic gardeners believe that interplanting species with poisonous or repellent qualities, such as marigolds, chrysanthemums, or garlic, with more palatable crops cuts down on insect damage to the crops. This seems reasonable but it is not a topic that the agricultural colleges have done much research on.

Probably no protective device is perfect, at least for very long. The evolutionary development of a protective poison by a plant presents an enormous opportunity for a herbivorous species if it can somehow evolve immunity; that is why there are tobacco worms. But for many herbivorous insects tobacco plants are unavailable even where there is a field full of them.

The prey that are available are those unable to defend themselves or run away, particularly for vertebrates preying on animals close to their own size. Such animals as wolves and mountain lions, consequently, tend to take the old, the sick, the weak, and the deviant. One result of this is that predation often is "prudent"; by taking animals that, on the average, are not going to live much longer anyway the predator seems to be managing the prey crop for sustained yield.

Within the range of items that animals are able to catch and use as food, some are usually preferred. Red foxes do not eat many apples if they can get white-footed mice and they do not eat many white-footed mice if they can get meadow mice. When food is scarce an animal may take what it can get, but when food is abundant the animal tends to concentrate on its preferred food.

Starvation is a powerful selective force for the predator to improve its hunting, and being eaten is a similar selective force favoring better methods of escape by the prey. Predator and prey are in a kind of evolutionary race. If one wins the race, then we no longer see them as a predator-prey system. If the prey become so adept at escape that they are no longer taken, then the predator must turn to another food (or become extinct). If the predator becomes so efficient a hunter that it exterminates not only the old and sick but the young and healthy, then the prey is gone and so is the predator, again unless it can turn to other foods. Thus the predator-prey systems that we see are those that work, those in which the predator does not overeat its prey nor the prey starve its predator.

The occasional situation in which predator, parasite, or disease does virtually exterminate a prey or host is usually one in which the species are newly exposed to one another; they have not yet started an evolutionary race. Familiar examples are the almost complete annihilation of the American chestnut by the chestnut blight introduced from Asia around the turn of the century, the virtual elimination of American and slippery elm in much of the eastern U.S. by the Dutch elm disease brought to this country from Europe in the late 1920's, and the near extinction of the lake trout in the Great Lakes once the sea lamprey was allowed entrance (by the construction of the Welland Canal around Niagara Falls).

These occurrences, along with the success stories in biological control and some other evidence, make it clear that the numbers of an organism can be limited by predation—that is, predation can set the carrying capacity. Hunters, of course, have never had any doubts about this. Hawks eat quail; consequently, a hawk killed means more quail to be hunted. Game biologists and ecologists in general are much less certain that predation is, for very many species, a limiting factor. Paul Errington's research on bobwhite (and muskrats) convinced him that up to a certain density these animals were almost immune to predation, except as an occasional accident. This density was set, however, by food and cover and by the animal's perception of what was crowded and what was not. "Surplus" birds were forced from favorable habitat. Errington wrote:

> ...Although there can be fighting (including fighting between social groups that may not be dissimilar to human warfare), the limiting factor of social intolerance need not always take the form of overt antagonism or fighting. Some of the most significant intolerance can have such benign bird-between-bird manifestations as frictionless avoidance or withdrawals on the part of the individuals or groups that recognize their own superfluity in places where they do not belong.
>
> Even so, manifestations of social overpopulation can include plenty of trouble. Bobwhite equivalents of displaced persons wander in strange places or try to live in uninhabitable areas. In their wanderings, they tend to be harassed by and vulnerable to predatory attacks. Not only may they be vulnerable to such formidable predators as great horned owls and dashing blue-darter hawks and agile and clever foxes but also to rather weak and clumsy predators having no special aptitudes for preying upon grown bobwhites unless something goes wrong. To a considerable extent, it may not seem to make much difference what kills the birds that are trying to live under highly adverse, if not hopeless conditions. They are the have-nots and they do not need to have human intelligence to know it. They may lack food or cover; they may lack both; but, possibly as contributory as anything to their serious troubles, they lack what might be called the sense of rightness that enables them to do the best they can with what they have.*

A. J. Lotka and Vito Volterra, at the same time that they produced their useful models of competition and coexistence, produced another model dealing with predator-prey numbers. Unlike the competition model, this model is so oversimplified that it is not useful. It omits any idea of a carrying capacity for either species and any idea of a threshold density below which the prey is not vulnerable, and it is unrealistic in many other ways.

What has kept this model from simply being put away as an idea that did not work out is that it predicts cyclic changes in numbers of predator and prey. Cycles, or at least reasonably regular fluctuations in numbers of some kinds of organisms, do seem to occur, and the explanation of these has been a long-standing ecological problem. The most frequently recognized cycles are those of 3 or 4 years and 9 or 10 years. Animals that seem to show 3- to 4-year cycles include the snowy owl, the Arctic fox,

---

*From Paul L. Errington, "Of man and the lower animals," *Yale Rev.*, 51:370-383, 1962.

and several species of lemmings preyed on by these two predators. Animals showing apparent 9- to 10-year cycles include the snowshoe rabbit and the lynx (prey and predator) and various other animals such as the muskrat.

Exactly what causes these and other oscillations in numbers is unknown. They may result from more or less regular changes in the physical environment. The pioneer animal ecologist of the University of Illinois V. E. Shelford argued, for example, that variations in ultraviolet radiation received from the sun were important. It is probable that in many cases predator-prey interactions are important but, if so, the cycles do not seem to be the cycles of the Lotka-Volterra equations; they are more complicated than that.

FUNCTIONAL AND NUMERICAL RESPONSES AND THE CONTROL OF PREY NUMBERS BY PREDATORS

Predators can regulate the population size of their prey only if they can eat (or at least kill) more of the prey as the prey population size (density) increases. More precisely, the predators must eat a higher *percentage* of the prey population as the prey density increases since only then will predation be acting as a density-dependent factor on the prey.

A population of predators can eat more prey in two ways: (1) each individual predator can eat more; and (2) the number of predators can increase. The second way would occur by either immigration or reproduction. The change in feeding rate with prey density has been called the *functional response* (by the English ecologist M. E. Solomon) and the increase in predator numbers termed the *numerical response.*

C. S. Holling used Solomon's terms in a study of predation in Canada by small mammals on an insect pest, the European pine sawfly. He found that the functional response was an increase in prey eaten with increased prey density, but that the increase levelled off. A functional response curve might look something like this:

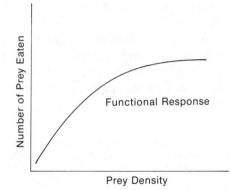

As prey density increases each predator eats more of the prey; however, at very high densities the number eaten only increases a little. The levelling-off occurs for various reasons; digging up, running down, peeling and pitting—or otherwise capturing and preparing the food—takes a certain irreducible amount of time. Also, animals get full or, in the technical term, *satiated*.

The numerical response may take a variety of forms, depending on whether an increase in prey density causes an increase in immigration, reproduction, or both or neither. In Holling's study one species of shrew showed no increase in numbers, but another species of shrew and a species of mouse did increase. The two kinds of numerical response curves, then, looked like this:

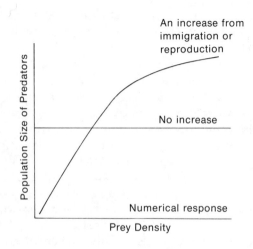

The combined response of any one species of predator would be the product of the functional and numerical responses. That is, if at low densities there were ten predators each eating ten prey per day, then 100 prey per day would be consumed. At high densities, if prey per predator increased to 20 and predator numbers also doubled, to 20, then 400 prey per day would be eaten.

Holling calculated the combined effect, functional and numerical, for all three species of small mammals preying on the pine sawfly and prepared another graph. The graph plotted the percentage of the prey population eaten by all three species against prey density. The curve (simplified a bit) looked like this:*

---

*This and the preceeding curves were adapted from C. S. Holling, "The components of predation as revealed by a study of small-mammal predation of the European pine sawfly," *Can. Entomol.*, 91:293–320, 1959.

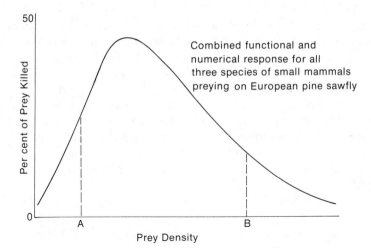

We may note that, since the percentage of the prey killed did increase over a wide range of densities, the small mammals of this ecosystem are acting in a density-dependent way and can regulate the population size of the species of insect. This could be true even for prey densities that are high enough to have gone past the peak of the curve (where the percentage killed is declining) as long as the percentage of insects killed is no less than the percentage added by insect reproduction. If, however, the insect is able to increase explosively from very low to very high densities, perhaps from A to B on the curve, then it can escape regulation by these predators.

MIMICRY AND OTHER KINDS OF
ADVANTAGEOUS RESEMBLANCE

Mimicry is the resemblance of one organism to another, not very closely related organism living in the same area. It was a puzzling phenomenon until the Darwin-Wallace theory of natural selection was put forth; then it, along with many other problems, became immediately comprehensible. H.W. Bates was the first (in 1861) to discuss mimicry specifically in terms of natural selection. He described the case in which some distasteful butterflies, not eaten by birds, were mimicked by another perfectly wholesome group of butterflies. He pointed out how a slight resemblance of some individuals of an edible form to a poisonous or otherwise distasteful form would lead to those individuals being passed over by predators, causing this kind of favorable variation to accumulate in later generations.

The explanation was greeted enthusiastically and in the next few years more cases of mimicry were described and similarly explained. Not much experimental work accompanied these descriptions and some persons grew disenchanted with the protective resemblance—the delusive similarity—explanation of mimicry. The naturalist and explorer William Beebe recounted an incident from some time early in this century when the great entomologist William Morton Wheeler, on his first trip to the tropics, visited Beebe in British Guiana. They were chatting and Wheeler told him "Beebe, in regard to mimicry, be sure to hold back; don't accept any new instances without complete evidence." Just then Wheeler's son, who had been exploring the area, brought in a tobacco can of half-dead insects and poured them out for his father to look at. Wheeler picked one up, looked at it carefully, and then abruptly threw it down. Then he looked at it again. "Beebe," he said, "believe anything you damned please about mimicry."

"He had," Beebe wrote*, "picked up a dead, yellow-banded, small-waisted wasp, which suddenly came to life and tried to sting him. On more careful scrutiny the insect proved to be a perfectly good moth, which in shape, proportions, antennae, pattern, color, and movement was an amazingly exact imitation of a wasp."

Since that time much experimental work has been done on mimicry and it has mainly supported Wheeler's second opinion; that is, it has supported the protective function of mimicry and its evolution through natural selection.

Mimicry in which an edible mimic resembles a distasteful or poisonous model has come to be known as **Batesian mimicry.** The example familiar to almost everyone is that of the monarch and viceroy butterflies. This system has been well studied by the husband and wife team of L. and J. Brower. In this case, the viceroy is the edible form which closely resembles the poisonous monarch. The poisons are compounds that produce nausea and vomiting in birds within a few minutes after being eaten or, in other words, fast enough for the birds to associate the effects with the food item. It is an interesting sidelight that the poisons seem to be manufactured by the plants (milkweeds) that the monarch feeds on, rather than by the monarchs themselves.

There are very many good examples of Batesian mimicry. There are harmless snakes banded like poisonous ones. In addition to moths there are flies and beetles which resemble various types of bees and wasps. Mimicry in plants has been described less frequently, but certain kinds of mints seem to be mimics of some of the stinging nettles.

Müllerian mimicry, first described by F. Müller in 1879, occurs when several poisonous or unpalatable species resemble one another. The success of Batesian mimicry depends upon the mimic staying fairly rare as compared to the model or else predators may tend to associate the pattern with edibility rather than inedibility. In Müllerian mimicry this problem does not exist; a predator eating either (or any, since often more than two

---

*In *The Book of Naturalists,* New York, Alfred A. Knopf, 1945, p. 250.

species are involved) of the Müllerian mimics will find it distasteful. A British biologist compared Batesian mimicry to an unscrupulous company copying the advertisement of a successful firm and Müllerian mimicry to several companies adopting the same advertisement to save money.

There are several other kinds of protective resemblance which resemble mimicry. There are many kinds of *cryptic* colorations or structures which act to camouflage the animal: there are insects shaped and colored like leaves and sticks; desert animals tend to be sand-colored; grassland birds have brownish, streaked backs; and there are as many more examples of this condition as there are of mimicry. On the other hand, many poisonous, stinging, or otherwise to be avoided organisms are brightly colored or strikingly marked, showing *warning coloration.* Examples are the black and white striping of the skunk and the black and orange striping of stinging insects. These are often the successful advertisers imitated by Batesian mimics.

Finally, there is *aggressive mimicry,* in which a predator or parasite has evolved a resemblance to its host or prey; the term has also been used if the predator is merely camouflaged (to look like part of a flower, for instance) or looks like some harmless species of the community. There are parasitic flies whose larvae feed on bee larvae; the adult flies resemble the bees that they parasitize and consequently are able to enter the hive to lay eggs without being attacked.

Another example of aggressive mimicry is seen in lightning bugs, or fireflies (which are really beetles). The flashings of lightning bugs are signals between the sexes. The males fly about on a summer night and produce flashes of light that are different for each species (depending on the species they may be bright or dim, long or short, fast or slow, and the lights may be slightly different colors). The females sit on plants and if they are ready to mate respond to the male by a flash that occurs at a characteristic interval (say, 5 seconds or 2.2 seconds) after the male's flash. The female continues to flash in response to the male's flashes as he homes in for mating. Large females of a different group of lightning bugs mimic the intervals of the females of the first group; the males are attracted to these females but when they arrive the males become not a mate but a meal.

INTERCOMPENSATION

Paul Errington has emphasized the interacting nature of factors affecting population size by calling them *intercompensatory.* Weather calamities may kill large numbers of organisms and, particularly at the edge of the geographic range of a species, may even destroy a local population. However, the effect is usually only a temporary (sometimes a very temporary) reduction of numbers. The individuals that survive have the best of habitat available to them, competition for food is low, and consequently they thrive and multiply. This is simply a restatement of how density-dependent population regulation works but it is worth emphasizing.

Hunters often believe that getting rid of the predators of some game species will mean more of that species available for them to shoot. This is sometimes true when predation is acting as a limiting factor. But it is often not the case, because saving the animals from predators results in the operation of other intercompensatory factors such as lowered reproductive success of the population.

Stocking programs are another way in which hunters and conservation departments, by not understanding intercompensation, have wasted money. If there are not as many pheasants, quail, or trout as the hunter or fisherman would like, they can be reared and "planted." But the way to raise the population level of a game animal is to raise the carrying capacity. Generally this means the difficult job of managing the habitat to provide a surplus of such essentials as good nesting and escape cover. Planted animals simply compete for food and cover with the ones already there; generally, as inexperienced animals set loose in a strange place, they quickly fall prey to local predators.

BIOLOGICAL CONTROL

We are emerging from a period in which an attempt was made to control pests almost solely through the use of chemical pesticides. *Biological control* generally refers to the use of predators or parasites of a pest to reduce its numbers to a point where it is no longer an economic problem. There have been several spectacular successes in biological control and a great many failures.

Although predaceous ants were used against citrus pests as early as the ninth century in China, most applications have been in the past hundred years. The earliest success in the U.S. was the control of the cottony-cushion scale in California, a small insect native to Australia that attacks citrus trees. Late in the 1800's, the vedalia beetle was imported from Australia and preyed on the scale insects so effectively that they were virtually eliminated from the citrus groves within two years. Since then, some 500 organisms have been imported into the U.S. and released for biological control purposes. About 100 became established and about 20 were successful in controlling the pest.

Another success was the control of klamath weed, which by the 1940's had become a pest in western U.S. grazing land. It invades overgrazed lands and is poisonous to cattle and sheep (and also to most insects because of a chemical called hypericin). Two leaf-eating beetles were imported from Europe; one of these was successful in nearly eliminating the plant from range land by keeping it defoliated and causing it to exhaust its food reserves. This case is interesting because it may shed light on the natural control of distribution and numbers of organisms. Klamath weed was not eliminated altogether. It still occurs in shady habitats where the beetles do not lay eggs, but not in open areas where it was formerly more abundant. If we studied klamath weed now without knowing its history, we might misinterpret its habitat preference and we would be unlikely to conclude that predation was limiting its abundance.

For many pests the predators or parasites that have been tried have evidently not been able to limit the pests' abundance. It may well be true that for some pests no organisms will have this ability, and this type of biological control will fail.

Most pests are, as we would expect, introduced species, freed from the ecological context in which they evolved. This is also the status of the control organisms. So far no organisms imported into this country for biological control have become important pests themselves, but the possibility is a consideration when using biological control.

The term "biological control" could logically include a much greater range of activities than is usually meant. For instance, the development and use of *resistant crop varieties* is biological. Increasing crop diversity in space and time by planting smaller fields and returning to crop rotation would probably help control some pests. Renewed attention to phenology (Chap. 4), as in delaying the planting of winter wheat until after the emergence of the Hessian fly, could be another type of biological control. Releasing large numbers of *sterilized males* into a population has proven effective in controlling the screwworm, a fly that lays eggs in the wounds of domestic animals. Males mate repeatedly but females only once, so that flooding an area with sterile males results in a large proportion of the females laying infertile eggs. Repelling pests by the use of companion plantings is already practiced by organic gardeners who interplant marigolds and garlic as protection for other crops (see p. 91). Bird pests may be repelled by playing tapes of their alarm calls through loudspeakers.

Alternatively, pest species can be attracted to traps by various means. Cowbirds are trapped on the breeding grounds of the Kirtland's warbler by placing a female cowbird in a trap. Her vocalizations, along with the food provided, attract other cowbirds which can then be removed from the warbler's breeding area. A method of attracting insect pests to their destruction is through the use of *pheromones,* chemicals used as signals between members of a species (see section on *Chemical Ecology* in Chap. 4). The use of a specific pheromone to attract a species of insect pest is promising and has the great advantage of being extremely selective.

A combined approach which includes biological control in the strict sense, additional methods such as those just described, and pesticides (Chap. 6) only when necessary has been referred to as *integrated pest control.*

HEALTH AND DISEASE

Health and *disease* are ecological terms which imply relationships between an organism and its environment. In elementary biology we usually learn the germ theory of disease in an oversimplified version: a particular bacteria or virus (a *pathogen*) enters the tissues of a host, multiplies, and produces a particular set of symptoms. This is true as far as it goes but, as René Dubos has pointed out in "Second Thoughts on the Germ Theory," the disease is an interaction of pathogen and host.* Most humans harbor microbes capable of producing disease but show no symptoms. The disease produced by one of the pathogens may be precipitated by some event that "lowers the resistance" of the individual—poor diet, excessive fatigue, or emotional stress.

One result of infectious disease (or sometimes merely exposure without the development of recognizable symptoms) is the development of immunity by the production of blood proteins, *antibodies,* that are specific for the particular pathogen. Because of the absence of specific immunities children may be more susceptible to infectious diseases than adults, and adults may be more susceptible to "new" diseases, perhaps imported from another locality.

An *epidemic* is the spread of a communicable pathogen through a population; its occurrence depends on the presence of a threshold number of individuals lacking immunity (called *susceptibles*), and its end depends on the decline of susceptibles (by immunity, death, or quarantine) to below the threshold. Some epidemics, such as plague, end with the population almost totally immune or dead and the disease organism locally extinct. For many pathogens, however, epidemics recur by the pathogen persisting until the number of susceptibles rises above the threshold density (generally through birth). Poliomyelitis and many childhood diseases showed such patterns before the widespread use of vaccines. Recurrent epidemics show many of the same features as predator-prey cycles as, in a specialized sense, they are.

Although we have focused on infectious disease there are several other kinds of conditions generally considered as disease, all with important ecological ramifications. There are deficiency diseases such as pellagra and scurvy, produced by improper diet which is, in itself, an environmental matter. This usually depends, in turn, on some sort of cultural dislocation relative to the environment; primitive peoples living on the diets which they have evolved rarely show deficiency diseases. It was not until corn was introduced into southern Europe and Africa that the burning, weakness, and eventual near-mummification caused by pellagra appeared. It was easier to grow enough calories using corn rather than other grains, but corn is deficient in niacin.

Sci. Am., 192:31–35, 1955.

There are inherited metabolic disorders such as diabetes and phenylketonuria. It is not known whether something in the environment causes the underlying metabolic problem, but the health of the individual *is* entirely an environmental matter. In environments where it is possible to obtain the hormone insulin in the case of diabetes and to reduce the amino acid phenylalanine in the case of phenylketonuria, the individual having these conditions may develop and live normally. In a similar fashion a hay fever sufferer may show no symptoms of allergy in an environment where ragweed does not grow.

Cancer may be considered as another disease category. In the past 20 years an enormous amount of money has been spent studying cellular and biochemical aspects of cancer. This research produced a great deal of basic scientific information on the functioning and growth of cells but not many useful methods for the prevention and cure of cancer. It now seems clear that the majority of cancers, possibly as much as 90%, are environmentally produced (Fig. 3–18). Cancer-producing (or *carcinogenic*) agents are various, from X-rays to cigarette smoke to food additives to viruses. Although it may be difficult to achieve, an environment allowing us to avoid carcinogens almost completely would practically eliminate the problem of cancer.

Stress-related diseases such as hypertension and gastric ulcers clearly have a strong environmental component. The environmental stress usually is some kind of social pressure. Many biologists have been impressed with the parallels in both physiology and behavior between crowded rats and mice and crowded people. Stressed humans and stressed rats both respond physiologically with such things as arteriosclerosis and ulcers. It is possible, although still not certain, that increases in crime, alcoholism, divorce, and mental illness in crowded inner cities are the same responses on the human level as are increased aggression, abnormal sexual behavior, and breakdown of normal social functioning in crowded mice.

J. M. May has described a situation in northern Viet Nam illustrating the importance of ecological factors in medicine and public health. In the delta region there people build their houses on the ground, with the stables on one side and an outdoor cooking area on the other. In the hills they build their houses on stilts, with animals kept in the space under the house and meals cooked inside the house. An anopheles mosquito which can carry malaria occurs in the hills but not in the delta region. It rarely flies higher than 8 or 9 feet above the ground, however, so that it usually encounters only animals. If it does fly as high as the living quarters the smoke from the cooking fire tends to repel it.

The rich delta lands are crowded. When people from the delta relocate in the hills they retain their delta culture and are consequently bitten by the mosquito and contract malaria. The delta people ascribe their troubles to demons in the hills. As we can see, they would be no more correct if

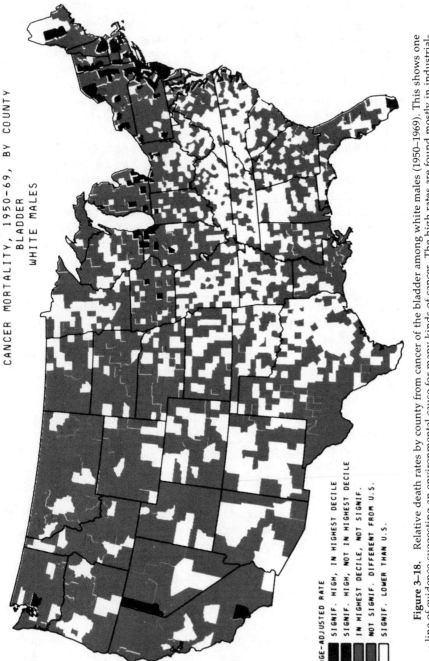

Figure 3–18. Relative death rates by county from cancer of the bladder among white males (1950–1969). This shows one line of evidence suggesting an environmental cause for many kinds of cancer. The high rates are found mostly in industrialized regions, especially those where industries produce or use organic chemicals. A map for women is similar but with less pronounced differences between the regions of high and low death rates; this suggests that exposure to carcinogens at work is especially important in causing bladder cancer. (From T. J. Mason, et al., "Atlas of cancer mortality for U.S. counties: 1950–1969," U.S. Department of Health, Education, and Welfare Publication Number (NIH) 75-780:18, 1975.)

they attributed their sickness solely to the malaria parasite and its mosquito carrier.

May concludes his article in this way: "The ancient formula of one ill, one pill, one bill, which seems to have been the credo of physicians for many generations should be abandoned. Disease is a biological expression of maladjustment. This is what should be taught to our students in medical school, and this phenomenon against which they are going to fight all their lives cannot be understood without an ecological study in depth [that includes] the environment, the host, and the culture."*

INTERSPECIFIC COMPETITION

COMPETITION AND THE COEXISTENCE OF SPECIES

With this topic, as in the sections on predation and disease, we move to the border zone between population ecology and community ecology. Here we are concerned with the interaction of populations of two different species which share a resource. Let us imagine that these are two birds sharing an insect food supply. We have defined competition as a combined demand in excess of the supply. In this case, whenever the combined demands of the two species exceed the supply of insects, whether by the birds being common or the insects being scarce, then these two species would be competing.

One possible outcome of such a situation is for one of the two species to become extinct. There is, in fact, a generalization to this effect known as *Gause's rule* (after the Russian ecologist G. F. Gause) or the *competitive exclusion principle* (a name proposed by Garrett Hardin). One form of Gause's rule makes use of the concept of the ecological niche: two species cannot occupy the same ecological niche. Incorporation of the niche concept (Chap. 4) is, however, unnecessary.

The baldest statement of Gause's rule is that complete competitors cannot coexist. This sounds reasonable; if we are willing to imagine two species that eat exactly the same thing, it seems likely that one of them will prove a little more efficient at getting the food or surviving starvation—or at something—that will give it the advantage and allow it to be around after the other is gone.

There are a few situations known in which one species seems to have replaced another or to have prevented the occurrence of another through competition. The moth skink (a type of lizard) became extinct in the Hawaiian Islands not long after another, closely related species was introduced. In central Illinois the presence of the black-capped chickadee in

---

*"The ecology of human disease," *Ann. N.Y. Acad. Sci.*, 84:789, 1960.

the forests along stream systems seems to prevent the Carolina chickadee, a very similar southern bird, from being able to nest north of the 39th parallel (Fig. 3–19). On the stream systems where no black-capped chickadees occur, Carolina chickadees nest north another 70 miles to 40° north latitude.

Much more common are cases in which one species seems to be excluded by another from a habitat. Where the willow tit and the marsh tit are *sympatric* (occur in the same geographical area), the willow tit lives in coniferous and mixed forest and the marsh tit in broad-leaved forest (Table 3–10). In a few regions only one species occurs. Where the willow tit is absent, the marsh tit occupies both coniferous and broad-leaved

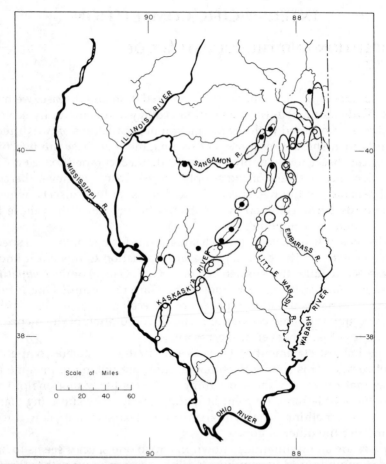

Figure 3–19. On the Embarass and Little Wabash Rivers, black-capped chickadees (*dots*) do not occur and Carolina chickadees (*circles*) extend all the way north, above 40°N latitude. On the Kaskaskia River, only about 20 miles west, black-capped chickadees occupy the northern section of the stream and Carolina chickadees extend north only to about 39°N latitude. (From R. Brewer, "Ecological and reproductive relationships of black-capped and Carolina chickadees," *Auk*, 80:11, 1963.)

Table 3-10 HABITAT DISTRIBUTION OF MARSH AND WILLOW TITS IN AREAS OF SYMPATRY AND ALLOPATRY

Habitat	Regions where Both Species Occur	Regions where only Marsh Tit Occurs	Regions where only Willow Tit Occurs
Coniferous and mixed forest	Willow tit	Marsh tit	Willow tit
Broad-leaved forest	Marsh tit	Marsh tit	—

woods. This suggests that the marsh tit's absence from coniferous forest in areas of sympatry is the result of competition with the willow tit. (Where the willow tit occurs alone it does not spread into broad-leaved forest, indicating that the habitat itself is somehow unsatisfactory.) Such a case may be considered competitive exclusion (from a habitat); however, it may just as logically be considered coexistence (within a geographical area).

A great deal of work has now been done in which laboratory ecologists have made complete competitors out of two species by confining them in cans or vials with a very simple environment and only a single kind of food (Fig. 3-20). Such work has been performed with protozoa and grain beetles, among other animals. One species always becomes extinct. These studies are valuable for showing that "competitive exclusion" really does occur. Two very similar species, each of which will live alone quite satisfactorily in a certain simple environment, may not be able to live there together.

The competitive exclusion principle may be satisfying in its simplicity, but it is not otherwise very satisfactory. Complete competitors must be rather rare, after all; the potentialities of the natural world are so vast that it is hard to imagine that the situation in which two species are *exactly* the same in their diet or nest sites is very common. The realistic question, then, is to ask just how strong competition has to be before coexistence is impossible.

Unfortunately, the general answer turns out to be one that is simple when expressed mathematically (as it was by A. J. Lotka and Vito Volterra in the mid-1920's) but rather difficult to state briefly and clearly in words. Roughly the answer is this: two competing species will be able to coexist if the effects of crowding are more severe intraspecifically than interspecifically. Where this is not the case, one of the two will be eliminated. For many cases, this is equivalent to saying that coexistence is possible where intraspecific competition is more important than interspecific competition. This is achieved principally by small differences in the use of resources that are limiting to the populations at one time or another. Two kinds of birds may, for example, forage for insects in slightly different places, say one on tree trunks and the other on branches. The trunk-feeding bird, when it is common, will deplete its own food supply while

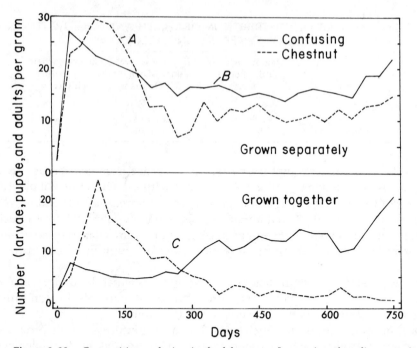

Figure 3–20. Competitive exclusion in the laboratory. In a series of studies spanning about 40 years, Thomas Park and his colleagues at the University of Chicago thoroughly investigated the population growth and competitive relations of two small grain beetles. They will be referred to here as the chestnut (chestnut-colored) grain beetle and the confusing grain beetle (confusing to early entomologists because it greatly resembles the chestnut grain beetle). In *A*, the chestnut grain beetle was grown separately in a culture consisting of 8 grams of flour and yeast, with the temperature maintained at 29.5°C (about 85°F) and relative humidity 60 to 70%. In *B*, the confusing grain beetle was grown separately under exactly the same conditions. In *C*, the two species were grown together in the same container. It can be seen that the chestnut grain beetle in *A*, living by itself, persists indefinitely (the graph covers more than two years). The same is true of the confusing grain beetle in *B*. But when the two are grown together, *C*, the chestnut grain beetle eventually becomes extinct. Extinction of one species was the invariable result of many trials with mixed cultures of the two species, but the species which became extinct varied according to environmental conditions. In a cool dry environment the confusing grain beetle always survived, but in a hot moist environment the chestnut grain beetle always survived. Under intermediate conditions the survivor was sometimes one species and sometimes the other, showing the importance of chance events whenever biological systems are considered, even under controlled laboratory conditions. (From H. G. Andrewartha and L. C. Birch, *The Distribution and Abundance of Animals*, Chicago, University of Chicago Press, 1954, p. 428.)

leaving much of the food supply of the branch-feeding species untouched. The same is true in reverse, for the branch-feeding species.

The differences in the use of resources sufficient to permit coexistence may be large, such as the restriction to different habitats in the case of the willow and marsh tits, or they may be very subtle. In the next section we will describe a situation in which two species of bees live off pollen from the same plant but manage to coexist by one concentrating on clumps of the plant and one concentrating on dispersed individuals of the plant.

Probably many of the differences which act now to limit interspecific competition and thereby to allow coexistence have evolved with the unfavorable effects of interspecific competition as the main selective force. As a simple example, suppose we have two species that overlap widely in the size of food they eat, and suppose also that the size of the food taken is at least partly based on hereditary factors (for example, the bird that takes slightly more small items has a slightly smaller bill). Interspecific competition will lead to poorer survival and reproduction among the individuals feeding on the medium-sized food items. Accordingly, natural selection will favor the individuals feeding on smaller items in the small-beaked species and individuals feeding on larger items in the large-beaked species (Fig. 3–21). If one species is more efficient, then the effects on survival and reproduction will be greatest on the other species. In this way the species diverge slightly and intraspecific competition becomes more important than interspecific competition; an accommodation is reached and the community can support both of them. This sort of thing has probably happened repeatedly in the history of all communities.

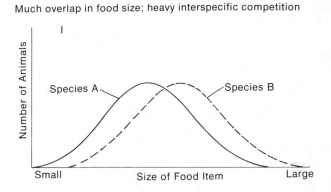

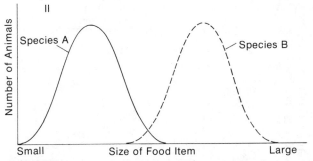

Figure 3–21. Natural selection reducing competition by reducing overlap in size of food items eaten; before (*I*) and after (*II*). The animals of species A eating small items and those of B eating large ones are more successful than individuals of either species eating medium-sized food; they have more offspring and live longer. Consequently, in the course of time, the "small" genotypes of A and the "large" genotypes of B increase and the diets diverge.

Competitive exclusion is what happens when, for some reason, competing species do not find the means for cutting down interspecific competition.

CAN TWO SPECIES LIVE ON ONE RESOURCE?

Whether there can be more species than resources depends entirely on what a resource is. Garbage is garbage until someone makes a profit producing methane from it, at which point it becomes a resource. It is certainly true that the resources supporting two species may be only very slightly different. One example comes from a study in tropical dry forests in Costa Rica by L. K. Johnson and S. P. Hubbell. They found that two species of stingless bees were able to coexist on the pollen of a single species of flowering shrub.

They observed that one species foraged in groups and gathered pollen from large clumps of the flowering shrub. The other foraged singly on scattered individual shrubs (plus small clumps). Johnson and Hubbell constructed the sort of diagram that economists use in the analysis of cost-benefit ratios. They assumed that feeding rate would go up and tend to level off, as plotted against resource density. (This is the same curve called a *functional response* curve in the earlier section on predation.) This, then, is the benefit derived from greater densities.

As resource density increased it was assumed that the costs of foraging would decrease. The costs are the amount of energy required to find the food source and to make use of it (including any energy used to defend it). Some measure of energy, such as kilocalories, would be appropriate for expressing both the benefits (for example, kilocalories obtained per day of foraging) and costs (kilocalories used per day). A graph combining the two curves would look like this:

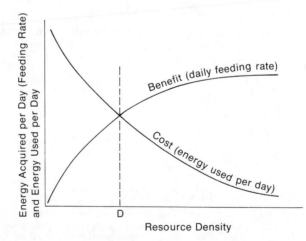

D is the break-even point. At any density below D the species could not make a living off the resource; it would have to spend more energy than it obtained. At any density above D it could make a living.

Plotting both species (A and B) on the same graph would look like this:

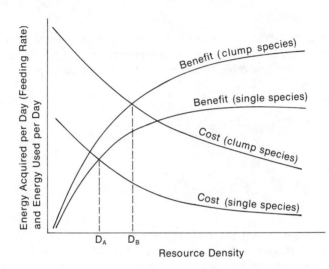

The species foraging on clumped resources (the "high density specialist") has the potential for increasing its feeding more with higher density for several possible reasons. Once in a large clump it would only have to spend a little time in travelling from plant to plant. Also, since it relies on scouts and then forages in groups at the site found by the scouts, less time may be spent per individual in locating clumps. This species also has higher energy costs, however, especially at low densities, because it must find and travel to clumps which at low densities will be far distant rather than making use of nearby individual plants. Also, for the model to work, the high-density specialist must defend clumps against the low density specialist, and this takes energy. Such defense was observed by Johnson and Hubbell.

The situation then is that the high density specialist uses clumps of the resources which it defends against the low density specialist. The low density specialist uses the isolated plants that the high density specialists cannot make a living on. The low density specialist could, of course, make profitable use of the clumps if it were not prevented from doing so by the other species.

MATHEMATICAL REPRESENTATION OF INTERSPECIFIC COMPETITION (THE LOTKA-VOLTERRA MODEL)

In an earlier section on the logistic curve we represented the growth of one population by

$$\frac{dN}{dt} = rN\left(\frac{K - N}{K}\right)$$

where N is population size, r is the intrinsic rate of natural increase, and K is carrying capacity. If two species are competing—if one is using some of the resources of the other—then we may rewrite the equations for the two species as follows.

For species 1:

$$\frac{dN_1}{dt} = r_1N_1\left(\frac{K_1 - N_1 - \alpha N_2}{K_1}\right)$$

For species 2:

$$\frac{dN_2}{dt} = r_2N_2\left(\frac{K_2 - N_2 - \beta N_1}{K_2}\right)$$

The only change is that for each species the term αN_2 or βN_1 has been added. The coefficients α and β are competition factors indicating respectively the effect of species 2 on 1 and the effect of 1 on 2. Note that the addition of each new individual of a species has an inhibitory effect on its further population growth. The inhibitory effect of one more individual of species 1 on itself is $1/K_1$; the inhibitory effect of one more individual of species 2 on itself is $1/K_2$. α and β are coefficients which express the inhibitory effects of each species in terms of the number of individuals of the species with which it is competing. The inhibitory effect of one new individual of species 2 on the growth of species 1 is α/K_1. For example, if two species are each competing for grass and species 2 eats three times as much as species 1, then α is 3. In other words, the inhibitory effect of species 1 on itself is $1/K_1$ but the inhibitory effect of species 2 on species 1 (that is, α/K_1) is $3/K_1$.

Any one species will stop growing when its carrying capacity has been reached by the *combination* of its own numbers plus the individuals of the other species (multiplied by the appropriate competition coefficient). That is, species 1 stops growing when

$$N_1 + \alpha N_2 = K_1$$

and species 2 stops growing when

$$N_2 + \beta N_1 = K_2$$

Notice that the two-species system will be at equilibrium only when the point at which growth stops coincides for the two species. Otherwise one species will keep growing, meaning that the other species will have to decline.

We can understand these ideas more easily using graphs. In the following graph the straight line consists of all the mixes of species 1 and species 2 that add up to K_1. When N_2 is 0, then N_1 is equal to K_1. When N_1 is 0, then there is a number of N_2 equal to K_1/α. The line represents all the combinations of numbers at which species 1 will stop growing. Note that if there is a mix of N_1 and N_2 such as at X, then N_1 must decrease (as shown by the arrow). Similarly, if there is a mix such as Y, then N_1 will increase (as shown by the arrow).

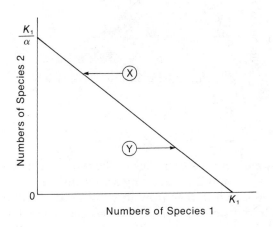

We can prepare another graph like the previous one but this time for species 2. It will look like this:

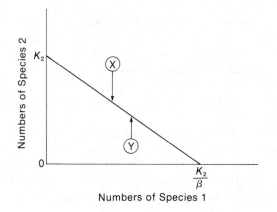

With a mix like that at X, species 2 must decline. With a mix such as that at Y, species 2 can increase.

We can now put the lines for the two species together on the same graph:

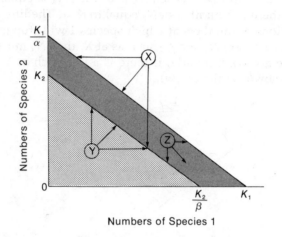

Numbers of Species 1

We can determine how the densities of the two species will change from points such as X, Y, and Z by drawing arrows for each species, as we did before, and then drawing the arrow in between that will be the resulting vector (just as in physics). Consequently, we can see that anywhere below line $K_2 - K_2/\beta$ (in other words, in the lighter area), both species will be able to increase their numbers. For example, if point Y represents 20 individuals of species 2 and 30 individuals of species 1, they might be able to grow to 40 of species 2 and 50 of species 1. Once the mix of species forms a point lying between the two lines (in the darker area), however, species 1 can continue to increase but species 2 must decline. Note that the combination of the two densities could move into this zone without any growth in numbers of species 2. If species 2 stays the same but species 1 grows in numbers, this will slide the point horizontally across the line, whereupon species 2 will have to begin to decline in numbers. In the clear zone above the line $K_1/\alpha - K_1$, species 1 as well as species 2 has to decline.

If you draw other arrows on the graph you will see that they will always end up in the same place. This is where $N_1 = K_1$ and $N_2 = 0$. In other words, the extinction of species 2 is the invariable result.

Three other relationships are possible in a graph of this sort (as long as we use straight lines). The second case is the opposite of the one just examined:

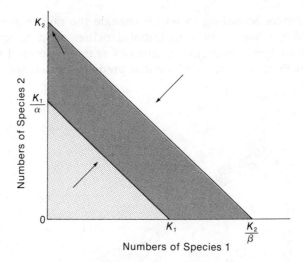

Numbers of Species 1

Here, since species 1 cannot increase past the mix of densities represented by the lighter area, and species 2 can increase in this area and also in the darker area, species 2 is the invariable winner, and species 1 becomes extinct.

In the third and fourth cases, the lines cross. Below is the third case:

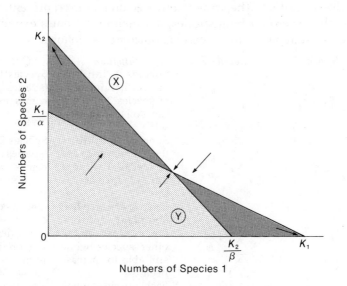

Numbers of Species 1

Here there are three possible equilibria. One of these is unstable; this is the one in the middle of the graph in which the two species coexist. Any event that shifts densities into the triangles will lead to the extinction of one species or the other. Which species wins and which becomes extinct, however, differs according to which triangle the mix of species enters. Consequently, if we begin with initial densities as in X, species 2 will win, but if we begin with initial densities as in Y, species 1 will win.

The fourth case is the only one that predicts coexistence:

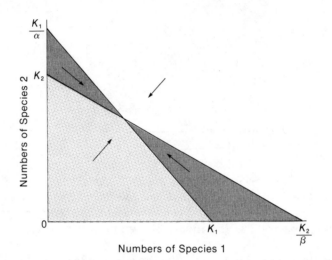

Here each species, as it becomes abundant, loses the capacity to increase before the other does. The arrows converge on a point representing some number above zero for both species, and neither becomes extinct.

We can summarize these various outcomes as follows:

Species 1	Species 2	Situation	Outcome
$K_1 > \dfrac{K_2}{\beta}$	$K_2 < \dfrac{K_1}{\alpha}$	Species 1 inhibits the further increase of species 2 while it can still increase itself	Species 1 wins (species 2 becomes extinct)
$K_1 < \dfrac{K_2}{\beta}$	$K_2 > \dfrac{K_1}{\alpha}$	Species 2 inhibits the further increase of species 1 while it can still increase itself	Species 2 wins (species 1 becomes extinct)
$K_1 > \dfrac{K_2}{\beta}$	$K_2 > \dfrac{K_1}{\alpha}$	Each species, when abundant, inhibits the increase of the other species while still able to increase itself	One species or the other wins, depending on initial numbers
$K_1 < \dfrac{K_2}{\beta}$	$K_2 < \dfrac{K_1}{\alpha}$	Each species, when abundant, inhibits its own further increase more than it inhibits the further increase of the other species	The two species coexist

As a final point, notice that although the outcome of interspecific competition (in this model) depends on the inhibitory effect of each species on the other, it also depends on the relative sizes of K. One species may have the advantage if we look only at the competition coefficients but may nevertheless lose if its K is very low relative to that of the other species. Suppose that species 1 has a competition coefficient (β) of 1.2 and species 2 a competition coefficient (α) of 0.8. This indicates that one individual of species 1 outcompetes one individual of species 2. But suppose that K for species 1 is 20 and K for species 2 is 80. The graph would look like this, and the outcome is clearly the extinction of species 1:

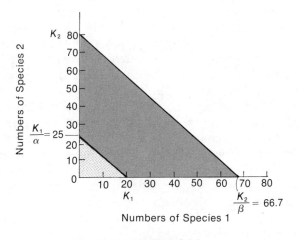

EXTINCTION

In the long run every species becomes extinct. The *extinction* may be merely technical; a line of evolutionary descent may be modified in the course of time so that it is species "z" now but was enough different 100,000 years ago that we would have to call it species "y". But most species of the past have come eventually to a definite end of the line; the last individual dies, the gene pool is gone. As the Cretaceous period of geological time drew to a close one species of dinosaur after another died out, until none was left. Two thirds of the families of the marine relatives of crustaceans called trilobites disappeared near the end of the Cambrian period; other species lived on for another 250 million years but none saw the end of the Permian.

To say that extinction occurs when death rate remains enough greater than birth rate that the population declines to zero may seem to say very little. It does, however, focus attention on extinction as a process of populations, different from other changes in population size only in its finality.

The basic cause of extinction appears to be a change in conditions causing increased mortality or decreased natality, for which the population cannot compensate. The change may be a new climate, a new predator, or a new competitor. If the species can adjust behaviorally or physiologically to the change, or if it can adapt genetically, then it can survive.

The "specialist" species, those with a narrow range of tolerance or adapted to a single food, are more prone to extinction than are the "generalist" species. This is because a change in conditions is more likely to move outside a narrow tolerance range than a broad one. Also, the evolutionary remaking of extreme specializations (such as saber teeth) is probably a slower process than remaking modest ones. Specialization is, of course, no crime as long as conditions do not change. As we have seen, competition theory suggests that ordinarily species must specialize, relative to one another, to avoid extinction from competitive exclusion.

Some specializations associated with extinctions have included restriction to a single habitat, restriction to a single food item, and adaptation to stable conditions. The latter often involves a reduction in biotic potential and an increase in body size (see the discussion of r and K selection earlier in this chapter, pp. 85-88).

Extinction, then, no less than evolution, is a natural event. Man's role in extinction has been to speed up its rate through his remarkable effectiveness in causing change. Some of the most important changes by which man has been involved in extinctions have been habitat changes, including the reduction and elimination of natural ecosystems, overhunting, and the introduction of predators and competitors.

Loss of habitat was probably a prime cause of the extinction of the heath hen and the ivory-billed woodpecker. Overhunting was probably an important agent in the disappearance of the passenger pigeon and the great auk. Many species of vertebrates known to have become extinct in modern times have been island forms—the moas, the Sandwich rail of Hawaii and several other flightless rails on other islands, the Oahu thrush, the Laysan apapane, the Tristan coot. Although habitat changes and overhunting were involved in the extinction of some island forms, most of them seem to have perished from predation or competition (through either exploitation or interference) by animals introduced by man. Some introductions were deliberate; in earlier times, sailors would release goats on islands so that there would be a source of meat and milk on their next voyage. Others have been unintentional, as in the spread of the Norway and black rat to islands throughout the world. Other animals introduced have been mongooses, cats, rabbits, and the familiar list of game mammals and birds.

The frequency of extinctions on islands suggests that local extinctions of populations may occur more often than has been generally realized. If a population becomes extinct in a nonisolated habitat it may be reestablished by dispersal from surrounding areas so rapidly that we have no inkling that an extinction occurred. Recolonization is slower on islands so that we detect the extinctions. For the many species that occur only on a single island (generally through the colonization of an island by mainland stock and subsequent evolution there) of course, no difference exists between local extinction and extinction of the species.

A good many species are less common than they were previously. The Federal government and several states have recognized this as an area of environmental concern. Basically three categories of these species are recognized. *Endangered* species are those in imminent danger of extinction throughout their range. Examples are the blue whale, black-footed ferret, Kirtland's warbler, California condor, whooping crane (Fig. 3–22), Colorado River squawfish, and the Texas blind salamander. *Threatened* species are ones that seem to be approaching endangered status, and *rare* species are ones that, because of their low numbers and possibly peculiarities of life history or habitat, should be closely monitored.

For many of the endangered species studies are now under way which may show how their extinction may be prevented. For the Kirtland's warbler, a small bird having a total population of about 400 birds and occurring only in a few counties of Michigan, studies have shown

Figure 3–22. The whooping crane, an example of an endangered species. Fewer than 100 individuals now exist but the current population has increased from 21 in 1953. The breeding range of the species is now restricted to Wood Buffalo Park in Canada and the wintering range to the Aransas Wildlife Refuge in Texas (where this photograph of Jo, an injured bird, was taken 15 May 1950 by L. H. Walkinshaw).

that they have a narrow habitat requirement, basically areas of jack pines about the size of fruit trees. They do not occur in very young nor very old stands of the pines. Also, their nesting success is lowered by parasitism of their nests by the brown-headed cowbird. Accordingly, management for the warbler has taken two forms: (1) burning off areas of jack pine plains that have grown up too much for the warblers, generating new habitat, and (2) trapping and killing cowbirds to reduce the number of warbler nests parasitized. It is as yet too early to judge whether these efforts will be successful in increasing the population size of the warbler.

Bibliography

Allee, W. C. *Animal Aggregations: A Study in General Sociology.* Chicago, University of Chicago Press, 1931.

Atsatt, P. R., and O'Dowd, D. J. "Plant defense guilds," *Science,*193:24–29, 1976.

Bates, H. W. "Contributions to an insect fauna of the Amazon Valley," *Trans. Linnean Soc. Lond.,* 23:495 ff, 1862.

Birch, L. C. "The intrinsic rate of natural increase of an insect population," *J. Anim. Ecol.,* 17:15–26, 1948.

———. "Experimental background to the study of the distribution of insects. II. The relation between innate capacity for increase in numbers and the abundance of three grain beetles in experimental populations," *Ecology,* 34:712–726, 1953.

Brewer, R. "Ecological and reproductive relationships of black-capped and Carolina chickadees," *Auk,* 80:9–47, 1963.

———, and Swander, L. "Life history factors affecting the intrinsic rate of natural increase of birds of the deciduous forest biome," *Wilson Bull.,* 89:211-232, 1977.

Brower, L. P. "Ecological chemistry," *Sci. Am.,* 220:22–29, 1969.

Brown, J. "Territorial behavior and population regulation in birds. *Wilson Bull.,* 81:293–329, 1969.

Christian, J. J., and Davis, D. E. "Endocrines, behavior, and population," *Science,* 146:1550–1560, 1964.

Coale, A. J. "The history of the human population," *Sci. Am.,* 231(3):41–51, 1974.

Cole, LaMont C. "Some features of random population cycles," *J. Wildl. Mgt.,* 18:2–24, 1954.

———. "The population consequences of life history phenomena," *Quart. Rev. Biol.,* 29:103–137, 1954.

Connell, Joseph H. "The influence of interspecific competition and other factors on the distribution of the barnacle, *Chthamalus stellatus,*" *Ecology,* 42:710–723, 1961.

Covich, A. P. "Ecological economics of seed consumption by *Peromyscus:* a graphical model of resource substitution," *Trans. Conn. Acad. Arts & Sci.,* 44:71–93, 1972.

Darwin, Charles. *On the Origin of Species.* London, John Murray, 1859.

Davidson, James and Andrewartha, H. G. "The influence of rainfall, evaporation, and atmospheric temperature on fluctuations in the size of a natural population of *Thrips imaginis* (Thysanoptera)," *J. Anim. Ecol.,* 17:200–222, 1948.

Deevey, Edward S., Jr. "Life tables for natural populations of animals," *Quart. Rev. Biol.,* 22:283–314, 1947.

Dumond, D. E. "The limitation of human population: a natural history," *Science,* 187:713–721, 1975.

Einarsen, A. S. "Some factors affecting ring-neck pheasant population density," *Murrelet,* 26:2–9, 39–44, 1945.

Elton, Charles. *Voles, Mice and Lemmings: Problems in Population Dynamics.* London, Oxford University Press, 1942.

Errington, Paul L. "Some contributions of a 15-year local study of the northern bobwhite to a knowledge of population phenomena," *Ecol. Monogr.,* 15:1–34, 1945.

———. "Predation and vertebrate populations," *Quart. Rev. Biol.,* 21:144–177, 221–245, 1946.

Fretwell, S. D. "Populations in a seasonal environment," *Princeton Univ. Press Monogr. in Pop. Biol.*, 5:1-217, 1972.

Gause, G. F. *The Struggle for Existence*. Baltimore, Williams & Wilkins, 1934.

Georghiou, G. P. "The evolution of resistance to pesticides," *Ann. Rev. Ecol. Syst.*, 3:133–168, 1972.

Greig-Smith, P. *Quantitative Plant Ecology*, 2nd ed. London, Butterworth, 1964.

Harper, John L. "Approaches to the study of plant competition," *Symp. Soc. Exp. Biol.*, No. 15:1–39, 1961.

Hensley, M. M., and Cope, J. B. "Further data on removal and repopulation of the breeding birds in a spruce-fir forest community," *Auk*, 68:483–493, 1951.

Hett, J. M., and Loucks, O. L. "Sugar maple (*Acer saccharum* Marsh.) seedling mortality," *J. Ecol.*, 59:507–520, 1971.

Huffaker, C. B. "Experimental studies on predation: dispersion factors and predator prey oscillations," *Hilgardia*, 27:343–383, 1958.

Johnson, L. K., and Hubbell, S. P. "Contrasting foraging strategies and coexistence of two bee species on a single resource," *Ecology*, 56:1398–1406, 1975.

Lack, David L. *The Natural Regulation of Animal Numbers*. New York, Oxford University Press, 1954.

Leslie, P. H. and Ranson, R. M. "The mortality, fertility, and rate of natural increase of the vole (*Microtus agrestis*) as observed in the laboratory," *J. Anim. Ecol.*, 9:27–52, 1940.

Lloyd, J. "Aggressive mimicry in Photuris: firefly femmes fatales," *Science*, 149:653–654, 1965.

Lotka, A. J. *Elements of Physical Biology*. Baltimore, Williams & Wilkins, 1925.

MacArthur, R. H., and Wilson, E. O. "The theory of island biogeography." *Princeton Univ. Press Monogr. in Pop. Biol.*, 1:1–203, 1967.

Malthus, T. R. *An Essay on the Principle of Population*. London, Johnson, 1798.

Miller, R. S. "Pattern and process in competition," *Adv. Ecol. Res.*, 4:1–74, 1967.

Nicholson, A. J. "An outline of the dynamics of animal populations," *Aust. J. Zool.*, 2:9–65, 1954.

Pearl, Raymond. *The Biology of Population Growth*. New York, Alfred A. Knopf, 1930.

———, and Reed, L. J. "On the rate of growth of the population of the United States since 1790 and its mathematical representation," *Proc. Natl. Acad. Sci.*, 6:275–288, 1920.

Pimentel, David. "Animal population regulation by the genetic feedback mechanism," *Amer. Nat.*, 95:65–79, 1961.

Rosenzweig, M. L., and MacArthur, R. H. "Graphical representation and stability conditions of predator-prey interactions," *Amer. Nat.*, 97:209–223, 1963.

Scott, T. G., and Klimstra, W. D. "Red foxes and a declining prey population," *South. Ill. Univ. Monogr.*, No. 1:1–123, 1955.

Slobodkin, L. B. *The Growth and Regulation of Animal Populations*. New York, Holt, Rinehart, and Winston, 1961.

Smith, J. S. "The role of biotic factors in the determination of population density," *J. Econ. Entomol.*, 28:873–898, 1935.

Solomon, M. E. "The natural control of animal populations," *J. Anim. Ecol.*, 18:1–35, 1949.

Van Valen, Leigh. "Life, death, and energy of a tree," *Biotropica*, 7:259–269, 1975.

Volterra, Vito. "Variations and fluctuations in the number of individuals in animal species living together," in R. N. Chapman, *Animal Ecology*. New York, McGraw-Hill, 1931, pp. 409–448.

Wangersky, P. J., and Cunningham, W. J. "Time lag in population models," *Cold Spring Harbor Symp. Quant. Biol.*, 22:329–338, 1957.

Wilson, E. O. *Sociobiology*. Cambridge, Harvard Univ. Press, 1975.

Wynne-Edwards, V. C. *Animal Dispersion in Relation to Social Behavior*. Edinburgh, Oliver and Boyd, 1962.

Young, Howard. "A consideration of insecticide effects on hypothetical avian populations," *Ecology*, 49:991–994, 1968.

Chapter 4

COMMUNITY AND ECOSYSTEM ECOLOGY: INTERACTIONS AND ORGANIZATION

TYPES OF INTERACTIONS

One of the first ecological observations to be recorded was that organisms live in *communities*. Certain species live together in a certain habitat and this combination of species tends to recur, more or less exactly, as the habitat recurs, with habitat and organisms bound together by interactions. This interacting system of community plus habitat is known as the *ecosystem*, a term first used by the English plant ecologist Sir Arthur Tansley.

F. E. Clements used three terms to describe the interactions between organisms and environment: *action, reaction,* and *coaction.* The environment *acts* upon the organisms of the community in many ways (see Chap. 2), including all the effects of temperature, wind, light, humidity, and soil moisture on the community.

The organisms of the community also *react* upon their environment. This meaning of the term *reaction,* the effects of organisms upon their physical environment, occurs only in ecology. Visualize a large area of bare dirt, perhaps a construction site, and then consider how it would be different if a forest, including not just trees but smaller plants, animals, and microorganisms, were suddenly there. The community would produce shade so that the light intensity would be lowered and temperatures would be moderated. The trees would act as barriers to wind and sound. As leaves fell and earthworms tunneled soil would be built. Plant leaves would intercept rainfall and allow it to reevaporate, and humus would soak up water so that runoff and erosion would be lessened. These effects are all reactions produced by a forest. The organisms of grasslands, lakes, and other communities produce their own reactions.

120

In an earlier section we defined *pollution* as the unfavorable modification of the environment as a by-product of man's activities. This is the same as saying that pollution consists of man's reactions. In this sense, pollution is a natural activity. However, man's reaction upon his environment has become a serious matter for the whole world for three reasons: (1) man has become a cosmopolitan species, occurring over the whole surface of the globe; no other species is so widely distributed; (2) humans are large both in physical size and in numbers; and (3) by utilizing energy subsidies, mainly from fossil fuel, man can exert effects (produce reactions) on the environment many times greater than a comparable animal which uses only the energy from its own metabolism.

The third kind of ecosystem interaction is *coaction*, the effect an organism has on another. One organism feeds on another and is in turn eaten by a third, setting up a food chain. An insect obtains food from a flower and thereby pollinates it. Rabbits crop off certain kinds of trees but leave black cherries, which grow up to dominate a field. Coactions may be as general as a tree limb serving as a site for a bird nest, or as specific as the relationship between the fig tree and its single pollinator, the Blastophaga wasp.

We have mentioned several types of coactions and discussed three, predation, disease, and competition (Chap. 3). Some other types of coactions are parasitism, commensalism, and mutualism.

Predation, Parasitism, and Disease. **Predators, parasites, and pathogens** all make a living at the expense of other organisms, their prey or hosts. Predators usually kill and eat their prey, which are about the same size or smaller than the predators. Parasites are generally small in comparison to their host and do not kill and eat it but obtain their food from body fluids or in some other nonfatal way (if the host is not too heavily infested and is otherwise healthy). Pathogens are microorganisms, usually bacteria or viruses, that sicken the host by living within it.

Anyone would identify a tiger, a tapeworm, and a polio virus as predator, parasite, and pathogen, respectively. However, there are interactions that are as much predation as parasitism and others that are as much parasitism as disease. Furthermore, there is no clear dividing line between parasites and disease organisms, on the one hand, and organisms that are a normal part of the body's flora and fauna, called *commensals*, on the other.

Commensalism. Commensals may be microorganisms such as the coliform bacteria that live in your intestine, slightly larger such as the mites that live in the oil glands around your nose, or ordinary-sized animals such as the house mice and house sparrows that live in or on your house rather than in or on you. Individuals of commensal species gain some advantage from the relationship; the effect is neutral for individuals of the other species. Other examples of commensalism not involving man are the use of prairie dog holes as nest sites for the burrowing owl and the use of old bird nests as nest sites for deer mice. A favorite textbook example is the remora, a fish with a suction cup on the top of its head, by which it attaches itself to a shark. It thus travels with the shark and eats the leftover scraps.

Mutualism. The dividing line between commensalism and mutualism is also unclear. Both individuals benefit from the association in mutualism. The usual example is the association of alga and fungus to form a lichen, but there are many others just as good. The relationship between termites which, like most organisms, cannot digest wood and the protozoans that live in their gut and can, is mutualism. So is the relationship between the rhinoceros and the tickbird, which rides around eating ticks and other insects off the rhinoceros.

The word *symbiosis* has sometimes been used to describe the relationship here termed mutualism. Symbiosis comes from roots meaning "to live together" so that, logically, it encompasses all coactions involving a continued, intimate association.

Action, reaction, and coaction are the bases for the structure and functioning of the community and the ecosystem. The structural and functional traits unique to this level of organization are the subjects of this and the following chapter.

ENERGY IN ECOSYSTEMS

TROPHIC STRUCTURE AND THE FOOD CHAIN

The word *trophic* means "feeding." The trophic structure of communities is based on the *food chain*, the sequence of organisms in which one organism feeds on the one preceding it. A typical food chain might be oak leaf → caterpillar → scarlet tanager → Cooper's hawk. In most communities several or many food chains exist which have interconnections at different points, forming a *food web*. The food web for most communities is very complex, involving hundreds or thousands of kinds of organisms (Fig. 4–1). One useful simplification is to group organisms into categories known as *trophic levels*, based on their position in the food chain (Fig. 4–2). The major categories are producers, consumers, and decomposers.

Producers. Producers (also called *autotrophs*) are organisms that can make food from simple inorganic materials. By *food* we mean complex organic compounds such as carbohydrates, fats, and proteins. Green plants are the producers with which most of us are familiar, and their food-making process is *photosynthesis*. In this process (see Chap. 2) plants use carbon dioxide, water, and some minerals, first, to produce carbohydrates and later various other organic materials, with oxygen being given off. Energy is as important as the materials involved. In the pho-

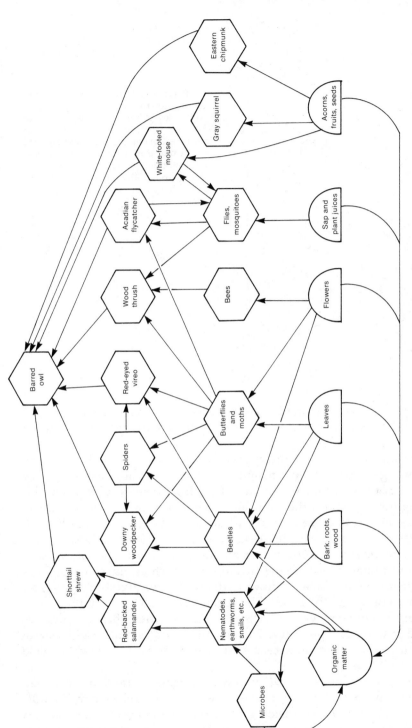

Figure 4-1. This food web for deciduous forest is greatly simplified as compared to the actual situation. Tens or hundreds of species have been lumped together in such categories as "beetles" and "nematodes, earthworms, snails, etc." Many species have been omitted, such as nuthatches, turtles, and fleas. Many connections are not shown. For example, mosquitoes bite other kinds of mammals in addition to mice, and other birds besides Acadian flycatchers eat mosquitoes. All of the animals provide food for the microbes. Despite these simplifications, the food web is still too complicated to comprehend easily. By grouping species into trophic categories, the situation becomes somewhat clearer.

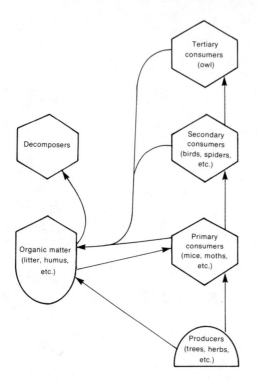

Figure 4–2. Here organisms of the deciduous forest are grouped into trophic levels based on the number of links that precede them in the food chain. All of the photosynthetic plants are in one trophic level, the producer level; all the organisms that eat mainly plants are in the primary consumer trophic level, etc.

tosynthetic process the radiant energy of sunlight is converted to chemical energy and stored in the chemical bonds of the compounds made by the plants.

Consumers. These are organisms that obtain their food by consuming other organisms. If they consume plants they are called *herbivores*, or *primary consumers*. If they obtain their food from green plants indirectly, by eating other animals, they are called *carnivores*, or *secondary* and *tertiary* consumers. All organisms other than autotrophs, those that have to get at least some of their foods prefabricated, are known as *heterotrophs*.

All animals are heterotrophs, as are fungi and most types of bacteria. We are most familiar with larger heterotrophs such as ourselves (or frogs or rats) that eat food, break it down partially in their digestive tracts, and absorb it mainly into the blood. The organic compounds from the bloodstream are absorbed by various cells in the body, where these compounds are used in two principal ways. They may be used as building blocks for other compounds or for new cells or they may be broken down to yield energy.

The latter process is *respiration,* in which organic compounds are combined with oxygen; the stored energy is released and carbon dioxide, water, and some mineral wastes are formed. The energy is that used by the organism for all its work—such as repairing and constructing cells, moving around, courting, fighting, and catching more food. The energy used for these functions is then given off as heat by the organism into the environment. Most of the carbon dioxide, water, and minerals resulting from respiration are *excreted* in one form or another.

Respiration is a universal process; every organism, heterotroph or autotroph, including green plants, respires. The energy that a plant uses for its work, such as growing and flowering, comes from the respiration of foods previously produced by photosynthesis.

Decomposers. These include scavengers and decay organisms such as fungi and many kinds of bacteria. They use dead plants and animals and excreta as their food source. The processes of digestion, respiration, and excretion are basically similar in all kinds of heterotrophs whether they are animals, bacteria, or fungi. Decay bacteria secrete their digestive enzymes into dead material outside their own bodies and absorb the food molecules rather than biting off a chunk and digesting it internally but these are really only details compared with the difference between autotrophy and heterotrophy.

If we go to the trouble to count the number of organisms in different trophic levels, we sometimes find that there are more plants than herbivores and more herbivores than carnivores, a pattern called the *pyramid of numbers*. A similar pattern, the *pyramid of biomass*, almost always results if dry weight is used instead of numbers. The pyramids of biomass

Figure 4–3. At the top are pyramids of numbers and of biomass (weight) such as might be expected in a thousand square meters of temperate grassland. C_1 = primary consumers, C_2 = secondary consumers, C_3 = tertiary consumers, and P = producers. There may be millions of grass plants, hundreds of thousands of grasshoppers, aphids, etc., thousands of carnivores such as spiders, and a few top carnivores like hawks or badgers. The pyramid in the middle represents the same situation but weight (specifically oven-dry weight) is used instead of numbers. The plants in a thousand square meters may weigh hundreds of thousands of grams, the primary consumers may weigh thousands of grams, etc. The third drawing, at the bottom, shows biomass by trophic level that does not form a pyramid. This sometimes occurs in open water of lakes or oceans where the producers are small and reproduce rapidly (single-celled algae, for example) and the consumers are large and long-lived (fish or large invertebrates).

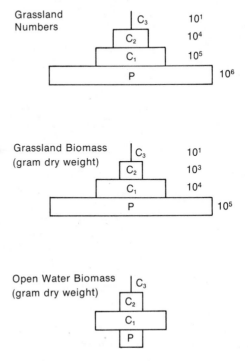

Grassland Numbers

C₃	10^1
C₂	10^4
C₁	10^5
P	10^6

Grassland Biomass (gram dry weight)

C₃	10^1
C₂	10^3
C₁	10^4
P	10^5

Open Water Biomass (gram dry weight)

C₃
C₂
C₁
P

and numbers are aspects of the structure of the ecosystem (Fig. 4–3). The functional basis of the pattern is in the flow of energy in the ecosystem.

ENERGY FLOW IN THE ECOSYSTEM

About 30% of the sunlight reaching the earth's atmosphere is reflected back into space, about 50% is absorbed as heat by ground, vegetation, or water, and about 20% is absorbed by the atmosphere, which does not seem to leave much for photosynthesis. In fact, only about 0.02% of the sunlight reaching the atmosphere is used in photosynthesis. Nevertheless, it is this small fraction on which all the organisms of the ecosystem depend.

Let us trace the path of energy in the community (Fig. 4–4). Of the energy stored in organic compounds in photosynthesis, the green plants themselves use some in respiration for their own growth and maintenance, with this energy being given off as heat to the surroundings. Primary consumers, or herbivores, eat plants. Some of the energy obtained in this way is stored in the growth of new tissue (and, in reproduction,

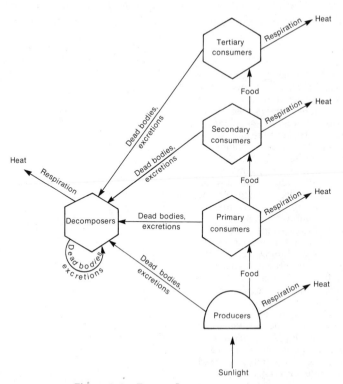

Figure 4–4. Energy flow in an ecosystem.

new individuals), but much is used for repair of tissues, moving around, and other activities that herbivores must perform to maintain themselves. Like the respiratory energy of plants, this energy is eventually converted to heat. Note that the herbivores have much less energy available to them than the plants originally produced in photosynthesis. The energy that the plants used in respiration is gone; also, some parts of the plant die before being eaten by herbivores and become food for decomposers, rather than for herbivores such as grasshoppers or deer. A good part of a plant leaf or the rest of a plant body is indigestible to most herbivores. These indigestible remains still contain energy and, by forming the major part of the feces of herbivores, become food for other decomposers.

The carnivores, or secondary consumers, obtain their energy from the herbivores in the same way that herbivores obtain theirs from plants. The carnivore uses some of the digested herbivore tissue for new cells, tissue, and producing young; the rest is respired to provide the energy for carrying on these activities. The energy available to the carnivore is, of course, much less than that taken in by the herbivore. Energy in materials that proved indigestible to the herbivore is gone and so is the large amount of energy used by the herbivores in their own maintenance. Also, like the primary consumer, the secondary consumer is not completely efficient in harvesting the available food nor in digesting what it does harvest.

Tertiary consumers feed on secondary consumers and quaternary consumers, if there are any in the ecosystem, feed on tertiary consumers. The foregoing processes occur at each of these levels (Fig. 4–5).

It is worthwhile to look quantitatively at the amount of energy lost between one trophic level and the next in the flow of energy in the ecosystem. Let us say that an average figure for the energy in sunlight striking one square meter of ground in the United States is about 1.5 million kilocalories per year (the actual figure will vary a bit depending on such factors as latitude and cloudiness). Only a very small fraction of this is stored in photosynthesis; we can use 1% as a very generous figure. Thus, about 15,000 kilocalories of the sunlight's energy are stored by plants in photosynthesis. The portion used by plants in their own maintenance may be from 15 to 75% of this amount, depending on the ecosystem. If 40% is considered as typical for plant respiration, then 60% of 15,000 is stored in new plant biomass. Consequently, the original 1.5 million kilocalories is now reduced to 9000 kilocalories potentially available to primary consumers.

In most ecosystems the energy in plant tissue is not used by grazers or browsers with high efficiency. Consumption by herbivores seems to vary over a wide range but is almost always less than 50% and is usually less than 20%. The rest of the plant material dies and goes to decomposer food chains. If we take 20% as another generous value, then about 1800 kilocalories of plant material are eaten by herbivores, or primary consumers. Of this amount, very roughly 10% is stored as new herbivore

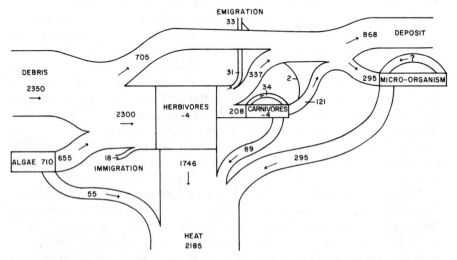

Figure 4–5. Energy flow for Root Spring, Concord, Massachusetts. Studying the energy flow through all the components of an ecosystem is an enormous task, and therefore most attempts involve many approximations. Many of the problems with precision were avoided in this study by dealing with a small, cold-water spring only 2 meters in diameter. The figures are kilocalories per square meter per year (figures in the herbivore and carnivore boxes are changes in standing crop). This ecosystem differs from the general example in the text in that import of energy in the form of organic matter, mainly leaves from surrounding trees, plays a large role. Not all of the energy entering is used, so that some goes into deposits on the bottom of the stream which will, in time, cause the spring to fill in (From J. M. Teal, "Community metabolism in a temperate cold spring," *Ecological Monographs*, 27(3):298, 1957. Copyright 1957 by the Ecological Society of America.)

tissue and is thus available to the secondary consumers, the first order carnivores. The other 90% is either used in maintenance of the herbivores and lost as heat or goes to decomposers as feces and excretions. Accordingly, about 180 of the 1800 kilocalories of plant material eaten by herbivores become new biomass.

If we assume that secondary consumers are slightly more efficient at harvesting the new biomass available to them, taking 30% instead of 20%, then 54 of the 180 kilocalories available in herbivores are ingested by first level carnivores. Using the same 10% efficiency figure that we used for the primary consumers for converting food to new biomass, these 54 kilocalories diminish to 5.4, which are stored as new carnivore protoplasm (Fig. 4–6). This amount is the energy potentially available to tertiary consumers. If we assume that they catch 30% and convert 10% of that, as we did for secondary consumers, then they ingest food containing 1.6 kilocalories and are able to produce new biomass containing 0.16 kilocalorie.

From this it should be plain why the pyramids of number and biomass occur. Less and less energy is available at each level. If a herbivore can find as much food as it needs in one acre, a secondary consumer of the same size will need ten acres and a tertiary consumer will need 100. This is one main reason, of course, why hawks and trout are relatively rare and mice and bluegill sunfish are common. It is also the reason why two people in a square mile of wilderness can live on trout but 2000 men in one square mile must eat rice.

Let us consider one more fact about decomposers. We have lumped under this one term all the organisms that gain their energy from dead bodies and excreta, but complicated food webs also exist among this group of organisms. Very small insects and other invertebrates may feed upon dead leaves that have fallen to the ground; these animals may in turn be eaten by small carnivorous invertebrates. Bacteria and fungi may grow on the remains of fallen leaves and the bacteria and fungi may be eaten by other invertebrates. The portion of the ecosystem which starts with dead protoplasm and is located mainly in the soil or in the bottom mud of bodies of water has not yet been well studied, but in many ecosystems, possibly most, the total amount of energy that flows in this way is greater than that which travels the more conspicuous route, from live plant to grazer to conspicuous carnivore (Fig. 4–7).

BIOMASS

In the previous section we discussed energy *flow*. At any one time each trophic level contains some amount of energy stored as biomass, often referred to as the *standing crop*. The pyramid of biomass gives an indication of the amount of energy present at a particular time. To understand the relationship between energy flow and the energy present as

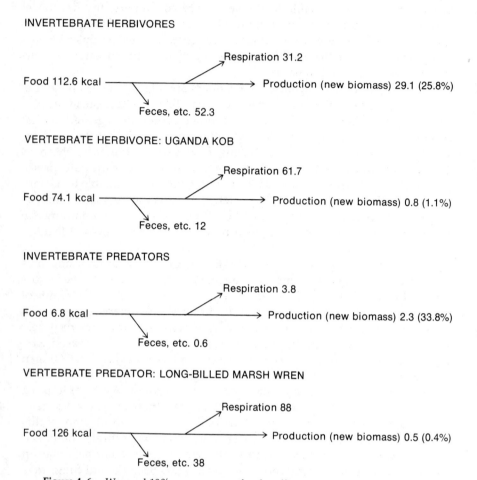

INVERTEBRATE HERBIVORES

Food 112.6 kcal → Respiration 31.2 → Production (new biomass) 29.1 (25.8%) → Feces, etc. 52.3

VERTEBRATE HERBIVORE: UGANDA KOB

Food 74.1 kcal → Respiration 61.7 → Production (new biomass) 0.8 (1.1%) → Feces, etc. 12

INVERTEBRATE PREDATORS

Food 6.8 kcal → Respiration 3.8 → Production (new biomass) 2.3 (33.8%) → Feces, etc. 0.6

VERTEBRATE PREDATOR: LONG-BILLED MARSH WREN

Food 126 kcal → Respiration 88 → Production (new biomass) 0.5 (0.4%) → Feces, etc. 38

Figure 4–6. We used 10% as an average for the efficiency with which consumers convert food into new protoplasm but the actual figures vary widely. One generalization that helps to make sense out of the variations is that homoiotherms are generally less efficient than poikilotherms. One reason is the substantial amount of energy homoiotherms must use just to maintain their body temperature. The diagram above compares energy flow through the invertebrate herbivores and predators (all poikilotherms) of a Tennessee grassland with energy flow through two populations of homoiotherms, an antelope in Africa and an insectivorous bird in a Georgia salt marsh. The figures are kilocalories per square meter per year, and the efficiency of conversion of foods to biomass is given in parentheses. Efficiencies vary from less than 1 to more than 30%, with the vertebrates (homoiotherms) being the low ones. (Data on: invertebrate herbivores and invertebrate predators from R. I. Van Hook, "Energy and nutrient dynamics of spider and orthopteran populations in a grassland ecosystem," *Ecological Monographs*, 41:20, 1971; vertebrate herbivore from H. K. Buechner and F. B. Golley, "Preliminary estimation of energy flow in Uganda kob," in K. Petrusewicz, ed., *Secondary Productivity of Terrestrial Ecosystems*, Polish Academy of Science, 1967, p. 252; vertebrate predator from H. W. Kale, "Ecology and bioenergetics of the long-billed marsh wren *Telmatodytes palustria griseus* (Brewster) in Georgia salt marshes," Nuttall Ornithological Club, Publication No. 5, 1965, p. 142.)

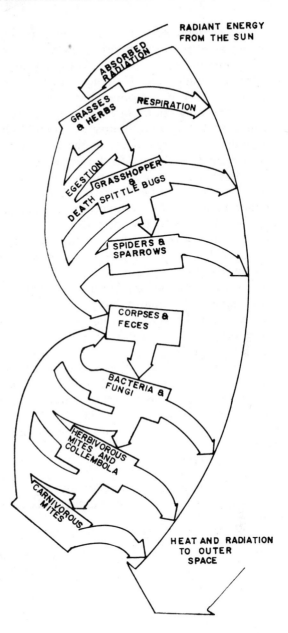

Figure 4–7. The importance of food chains that start with dead leaves, limbs, carrion, and "corpses and feces" in general is emphasized in this diagram of energy flow in an old field in Michigan. The food chains involving grasshoppers and sparrows are what we see when we look at a grassland but the food chains of bacteria and mites often process a greater proportion of the energy that the green plants have fixed. (From M. D. Engelmann, "The role of soil arthropods in the energetics of an old field community," *Ecological Monographs*, 31:235, 1961. Copyright 1961 by the Ecological Society of America.)

standing crop we can compare energy flow with cash flow in and out of your checking account. Suppose that your account has has $9.41 in it at the beginning of the year, and during the course of the year you put money in and take money out. If $500.00 is deposited and $500.00 taken out, then $9.41 still remains at the end of the year. The $9.41 is analogous to standing crop and the cash flow of $500.00 corresponds to energy flow.

Energy is deposited in an ecosystem through photosynthesis and withdrawn as respiratory energy at the various trophic levels. If as much energy goes in as comes out during a year, then the ecosystem is at an energetic *steady state*. If photosynthesis is greater than respiration in all the trophic levels combined (corresponding to putting $500.00 in the account and only taking out $400.00), energy must be accumulating somewhere in the system—either as larger organisms (for example, in tree growth), more organisms, or through storage in litter and humus. An ecosystem may also be on the negative side of a steady state; in a given year less energy may enter the system than is lost from it, in which case there is a net loss of energy from the system. The positive energy balance situation is probably common in early stages of community development (Chap. 5). The situation of negative energy balance probably cannot last for very many years but occurs under certain conditions such as severe drought. In a very dry year in a grassland much of the litter which has accumulated in preceding years may be decomposed and the year's production may be too small to make up the loss.

Table 4–1 compares energy flow in an immature forest (a pine plantation in England) and a mature one (a rain forest in Puerto Rico). The rain forest is obviously much more productive but the object of the comparison is to show that the rain forest is at an energetic steady state whereas the pine plantation is not. In the rain forest the sum of the energy used in respiration by the plants, animals, and bacteria equals the energy fixed in photosynthesis. In the pine plantation there is more energy fixed in photosynthesis than is used. The excess is *net community production* which accumulates in the form of wood in tree trunks, litter, etc.

PRODUCTIVITY

The total energy storage by autotrophs in an ecosystem is referred to as *gross production*. The plants themselves use a considerable amount of

Table 4–1 ENERGY FLOW IN TWO FORESTS*

	Immature Forest (Kcal/m²/year)	Mature Forest (Kcal/m²/year)
Total photosynthesis (gross primary production, GPP)	12,200	45,000
Plant respiration	4,700	32,000
New plant tissue (net primary production, NPP)	7,500	13,000
Heterotrophic respiration†	4,600	13,000
Net community production (NCP)	2,900	Little or none
Ratio NCP/GPP	23.8%	0%

*Modified from Odum, E. P., *Fundamentals of Ecology*, Philadelphia, W. B. Saunders, 1971, Table 3–5, p. 46.

†Includes herbivores, carnivores, bacteria, and other decomposers.

this in their own respiration. The amount of stored energy left after the plants' respiration is *net production*.* A year is a convenient interval to use in stating productivity, but daily or growing season production may also be studied. Energy units and weight are two different ways of expressing productivity. Either may be used because the energy is stored in the organic compounds which compose plant bodies. Dry weights are taken because the water content of different kinds of organisms varies.

The energy content of a given weight of plant or animal material is calculated by burning the dry material in a calorimeter and determining the amount of heat given off. This has been done for various kinds of plant and animal materials. The amount of heat given off by plant material runs about 4-4½ kilocalories per gram of oven-dried plant matter; animal material is usually 5-5½ kilocalories per gram of dry weight. Using these figures (which will be only approximate for a particular type of plant or animal) we can convert from weight to energy and vice versa.

Annual net production is a convenient basis for comparing various ecosystems because it is the energy potentially available to the organisms of the trophic levels past the producers for the year. Net yearly production ranges from zero in the driest deserts and other habitats too extreme to support plants to greater than 5000 grams per square meter (Table 4–2). This latter value, upwards of 11 pounds of plant material added to one square meter in one year, is a sizable amount of growth. Annual net production is probably between 500 and 2000 grams per square meter over much of the temperate part of the earth. On a pounds per acre basis, a familiar way of expressing agricultural production in English-speaking countries, this is about 5,000 to 18,000 pounds per acre. These are not actually *yields* in the agricultural sense since they include all new protoplasm including stems, bark, roots, and pollen. Humans could obtain this amount of plant material as food only if they were able to eat all parts of the plant and even then only if they were able and willing to eliminate all other animals that might also consume parts of the plants.

What environmental factors contribute to high annual productivity? A good moisture supply, a long growing season, warm temperatures, and high fertility are all favorable. It is not surprising that marshes and estuaries where moisture is abundant and nutrients are supplied by runoff from the land have a high rate of production. Nor is it surprising that tropical ecosystems with a long growing season and high temperatures have a high annual production and that deserts with their low rainfall and high evaporation rate are unproductive. However, it is interesting to note that the productivity of the open ocean is scarcely more than that of deserts. The main reason appears to be a scarcity of chemical nutrients—phosphates, nitrates, or possibly iron. The only really productive

---

*Sometimes these are called gross and net *primary* production, with energy storage at consumer levels of the ecosystem then referred to as *secondary* production.

Table 4-2 NET PRIMARY PRODUCTION AND PLANT BIOMASS (DRY WEIGHT) FOR THE EARTH*

This table is an attempt (by R. H. Whittaker and G. E. Likens) to summarize information on the biomass and productivity of particular ecosystems and then, using estimates of the earth's surface occupied by the various types of ecosystems, to estimate global totals. Figures for world net primary production are in billions of metric tons (one metric ton is one million grams, or about 2200 pounds).

Ecosystem Type	Area 10⁶ km²	Net Primary Productivity, per Unit Area g/m²/yr Normal Range	Mean	World Net Primary Production 10⁹ t/yr	Biomass per Unit Area kg/m² Normal Range	Mean	World Biomass 10⁹ t
Tropical rain forest	17.0	1000-3500	2200	37.4	6-80	45	765
Tropical seasonal forest	7.5	1000-2500	1600	12.0	6-60	35	260
Temperate evergreen forest	5.0	600-2500	1300	6.5	6-200	35	175
Temperate deciduous forest	7.0	600-2500	1200	8.4	6-60	30	210
Boreal forest	12.0	400-2000	800	9.6	6-40	20	240
Woodland and shrubland	8.5	250-1200	700	6.0	2-20	6	50
Savanna	15.0	200-2000	900	13.5	0.2-15	4	60
Temperate grassland	9.0	200-1500	600	5.4	0.2-5	1.6	14
Tundra and alpine	8.0	10-400	140	1.1	0.1-3	0.6	5
Desert and semidesert scrub	18.0	10-250	90	1.6	0.1-4	0.7	13
Extreme desert, rock, sand, and ice	24.0	0-10	3	0.07	0-0.2	0.02	0.5
Cultivated land	14.0	100-3500	650	9.1	0.4-12	1	14
Swamp and marsh	2.0	800-3500	2000	4.0	3-50	15	30
Lake and stream	2.0	100-1500	250	0.5	0-0.1	0.02	0.05
Total continental	149		773	115		12.3	1837
Open ocean	332.0	2-400	125	41.5	0-0.005	0.003	1.0
Upwelling zones	0.4	400-1000	500	0.2	0.005-0.1	0.02	0.008
Continental shelf	26.6	200-600	360	9.6	0.001-0.04	0.01	0.27
Algal beds and reefs	0.6	500-4000	2500	1.6	0.04-4	2	1.2
Estuaries	1.4	200-3500	1500	2.1	0.01-6	1	1.4
Total marine	361		152	55.0		0.01	3.9
Full total	510		333	170		3.6	1841

*From Whittaker, R. H., *Communities and Ecosystems*, 2nd ed., New York, Macmillan, 1975, p. 224. Copyright 1975, The Macmillan Company.

areas of the ocean are estuaries and coral reefs, where production may be as high as anywhere on earth. Areas of upwelling from the ocean floor (mainly along the western coasts of the continents) and other coastal areas where nutrients are carried out from rivers and estuaries have reasonably high productivity, similar to that of most terrestrial areas. Much of the open ocean, making up about 60% of the earth's surface, has a net productivity of 500 grams per square meter per year or less.

ENERGY IN AGRICULTURE

The basics of energy flow in agriculture are clear. Sunlight strikes a field on which man has planted a crop and from which he has eliminated competing vegetation. Some of the sun's energy is fixed in photosynthesis and stored in new protoplasm. The energy in the edible portions of the plant is then passed on to man. When man eats meat another step is involved, with cattle, pigs, or chickens eating the plants and man obtaining the energy stored in the new animal protoplasm.

This picture of energy flow is shown in Figure 4–8. The figures are growing season approximations for one acre of Illinois cornfield producing 100 bushels of corn an acre (the national average was just over 80 bushels/acre in 1970). Of the sunlight striking the acre (2043 million kcal), 1.6% is used in photosynthesis. Corn has the efficient C_4 pathway so that relatively little energy is used in plant respiration; 80% of the 33 million kcal gross production winds up as net production. Of this total production, only the grain is used directly as food. If man functions as a primary consumer, then approximately 8.2 million kcal is potentially available, disregarding indigestible parts of the grain, processing waste, and so forth.

If the corn is instead fed to livestock, placing man at the carnivore level, there are losses in such forms as material wasted or not digested by the livestock, respiration by the animals, and processing waste in the slaughterhouse. The result is that the human food in the form of beef represents about 5% of the kilocalories of grain fed. If the grain is fed to chickens or to pigs the loss is less, with human food in the form of poultry or pork containing about 10% of the energy in the grain. By American and western European dietary standards, a human needs about one million kilocalories per year, meaning that one acre of corn can feed about eight persons (remember, however, that 100 bushels per acre is well above the average even in the U.S.). When corn is converted to beef it provides only enough food for one person for about five months.

These figures emphasize the reason why high populations of humans must be mainly vegetarian (herbivorous) but it is not necessarily an argument against raising livestock. Meat may well be a requirement for optimal human growth and development, although not in the quantities

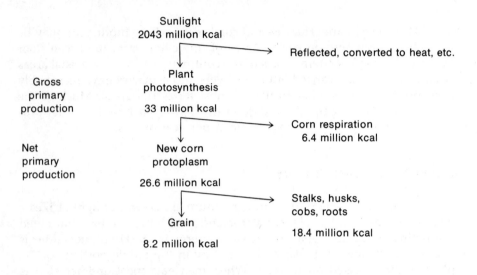

The grain may be consumed in some form by man. If, instead, it is fed to cattle to produce beef, the energy flow continues:

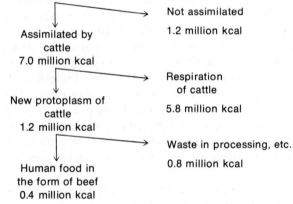

Figure 4–8. Energy flow in a corn field and a feed lot. (From E. N. Transeau, "The accumulation of energy by plants," *Ohio Journal of Science*, 26:7, 1926; G. H. Heichel, "Agricultural production and energy sources," *American Scientist*, 64:66-67, 1976.

that Americans consume. Much of the world's cattle production is on rangeland that for one reason or another, such as low or erratic rainfall, should not be cultivated. Ruminants such as cattle can live on the grass produced, whereas men cannot. Pigs and chickens are often raised on scraps and garbage although this is no longer true of commercial pork and poultry production. Also, as we shall see, low intensity meat production is as efficient (comparing food calorie output with petroleum calorie input) as is most high intensity vegetable farming (tomatoes and lettuce, for example) and fruit production (grapefruits and pears).

In the energy flow diagram for corn and beef the energy subsidy to agriculture is left out. In earlier times this subsidy was derived mainly from the metabolic energy of humans and draft animals performing such tasks as clearing, planting, and cultivating land. Now this subsidy is mainly in the form of petroleum, used for running machinery, manufacturing fertilizer and pesticides, irrigating fields, and drying harvested crops. David Pimentel of the New York State College of Agriculture and his coworkers calculated that nearly 3 million kcal of *"cultural"* energy go into the production of one acre of corn yielding, as we already showed, about 8 million kcal of grain. This 3 million kcal is in addition to the photosynthetic input of about 33 million kcal. The breakdown of the 3 million kcal by category is shown in Table 4–3.

In raising corn we get back about 3 kcal in grain for every 1 kcal spent as petroleum. Sorghum and sugarcane seem to be somewhat more efficient than corn, with wheat, oats, and soybeans being somewhat less efficient. All of these, however, return more than 1 calorie of food for each calorie of energy subsidy, which is not the case for several of the intensively grown fruit and vegetable crops. For peaches, pears, grapefruits, lemons, green beans, lettuce, melons, tomatoes, celery, and various other crops, more energy is put into the raising of the crop than is gotten out as food. Obviously if man tried to live on such crops and had only the energy of his own metabolism he would starve to death.

Even this does not tell the whole story. The energy used in food processing, the production of cans, jars, wrappers, and packages, the transportation of crops to the factory and processed food to grocery stores, and energy for refrigeration and cooking of food must be added to the

Table 4–3 CULTURAL ENERGY INPUTS IN CORN PRODUCTION*
Figures are kilocalories per acre. 1970 represents approximately current practice; 1945 is included to represent somewhat lower (although still high) technology and energy subsidies. Corn yield increased 138% in the 25-year span but the energy subsidy increased 213% with the result that return/input ratio was considerably lower in 1970.

Input	1945	1970
Labor	12,500	4,900
Machinery	180,000	420,000
Gasoline	543,400	797,000
Nitrogen	58,800	940,800
Phosphorus	10,600	47,100
Potassium	5,200	68,000
Seeds for planting	34,000	63,000
Irrigation	19,000	34,000
Insecticides	0	11,000
Herbicides	0	11,000
Drying	10,000	120,000
Electricity	32,000	310,000
Transportation	20,000	70,000
Total inputs	925,500	2,896,800
Corn yield (output)	3,427,200	8,164,800
Kcal return/input kcal	3.70	2.82

*From Pimentel, David, et al., "Food production and the energy crisis," *Science*, 182:445, 1973.

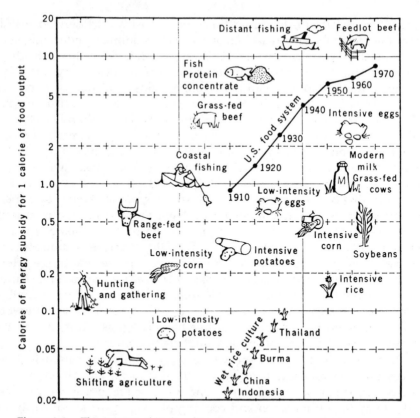

Figure 4–9. The amount of "cultural" energy subsidy various food crops require. At 1.0, each calorie of food output requires the input of a calorie (in addition to the energy of photosynthesis). Below this line the crops return more energy in food than they cost in petroleum or human labor; above this line they cost more energy than they return. (From C. E. Steinhart and J. S. Steinhart, *Energy: Sources, Use, and Role in Human Affairs*, North Scituate, Mass., Duxbury Press, 1974.)

energetic cost of crop production. When J. S. and C. E. Steinhart took this approach they found that the U.S. food system as a whole had passed the break-even point sometime between 1910 and 1920, and that by 1970 close to 10 calories of cultural energy were being expended for every calorie of energy on our plates. Figure 4–9 shows the energetic efficiency of the American food system and of the production of various specific food crops.

BIOGEOCHEMISTRY

NUTRIENT CYCLING

The energy that we have been discussing is in the form of chemical bonds of organic compounds made up of such elements as carbon, hydrogen, oxygen, nitrogen, sulfur, and about 20 (or probably more) others —the essential elements of protoplasm. These elements move from the abiotic to the biotic portion of the ecosystem as plants take in carbon dioxide from the air and water and minerals from the soil and use these to produce carbohydrates, fats, and proteins. As the consumers eat the plants and obtain the organic compounds of which the plant body is made these elements are passed on to the consumers. They are returned to the abiotic portion of the ecosystem through excretory processes of producers, consumers, and decomposers. This, then, is another aspect of ecosystem function—the flow of materials. Energy flow, as we have already seen, is a one way process. Energy enters the ecosystem as sunlight and leaves as heat which is dissipated in the universe. However, materials (matter) move in more or less complete circular systems known as *biogeochemical cycles*. This is probably an unnecessarily long name but it indicates that both organisms (*bio*) and the rest of the earth (*geo*) are involved in the cycling of these essential chemical elements.

The Carbon Cycle. Everyone is familiar with biogeochemical cycling in at least a vague way, through the processes of photosynthesis and respiration. Figure 4–10 illustrates the **carbon cycle**. Carbon dioxide

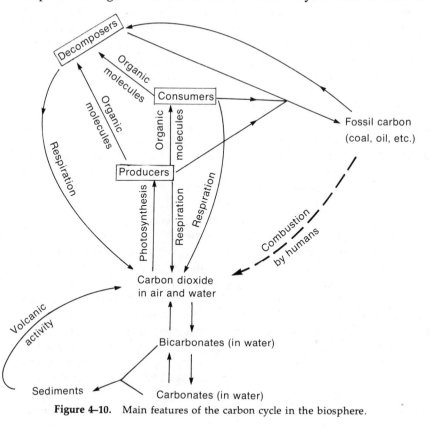

Figure 4–10. Main features of the carbon cycle in the biosphere.

in the air (or, for aquatic plants, dissolved in the water) is taken up by plants and used in photosynthesis. As organic compounds are used by the plants themselves some carbon dioxide is returned to the environment but much of the carbon is retained in the plant bodies. Primary consumers obtain their carbon when they eat plants and higher level consumers when they use lower level consumers as food. Decomposers act on the dead bodies of plants and animals when they die and obtain their carbon from these. All consumers and decomposers carry on respiration which returns carbon dioxide to air or water.

This is the basic part of the cycle but there are some further ramifications. There is a rather complicated chemical system involving carbon dioxide in water. When carbon dioxide dissolves some of it combines with water to form carbonic acid. Bicarbonates and carbonates may, in turn, be formed. Carbonates are not very soluble and may precipitate and be deposited as sediments on the bottom of lakes and oceans. All of these are reversible reactions; the effect of this is to buffer the carbon dioxide content of the air. If there is a local depletion of carbon dioxide in the air a reversal of the reactions just described occurs, resulting in the liberation of carbon dioxide from water to air. If there is an increase in carbon dioxide in the air more dissolves in water, more is converted to carbonic acid, and so forth. Material which becomes incorporated in sediments probably returns to the system slowly, but these sediments may be brought back into circulation by geological processes such as volcanic activity or by the uplift of land and subsequent weathering of limestone (calcium carbonate).

Another complication is the storage of fossil carbon. Some dead organic matter such as the peat in bogs escapes decomposition. The amount of new organic material formed without subsequently decomposing seems to be small but a fairly large amount has accumulated throughout the period of life on earth, much of it over a period of about 65 million years ending 280 million years ago. This period, the Carboniferous, was the time during which most of our coal, oil, and gas was stored. Before the advent of man this material probably decomposed slowly. There are, for example, petroleum bacteria that are able to use oil as their carbon and energy source (that is, as their food). With man's discovery of the use of this material as an energy source the release of the stored carbon as carbon dioxide has greatly increased. In 1900, the addition of carbon dioxide through burning of these fossil fuels to the atmosphere by man was about 1 billion tons, in 1940 it was about 2 billion tons, and in 1970 it was about 5 billion tons, enough to raise the carbon dioxide content of the air by 2.3 parts per million each year if none were removed. The carbon dioxide content of the air has increased but not by this amount; instead, between 1959 and 1969, it increased by about 1 part per million per year. Where has the rest gone? Presumably most of it has gone to the buffering system in the oceans. The carbon dioxide content of the air has nevertheless increased so that the buffering system has lagged behind the changes in the air.

The Nitrogen Cycle. The basic features of the *nitrogen cycle* are shown in Figure 4–11. This appears to be, and is, a complicated system. Let us begin with the nitrogen uptake of green plants. Note that the large reservoir of nitrogen in the air is not usable by green plants, most of which must take in their nitrogen in the form of nitrate. This they use mainly in the production of proteins. Animals produce their own proteins from plant proteins in their food, and decomposers make their proteins from those in dead animal and plant tissue.

When proteins are broken down in respiration a waste product containing nitrogen is produced. The usual waste product of the cell is ammonia. Some organisms excrete ammonia directly; however, ammonia is

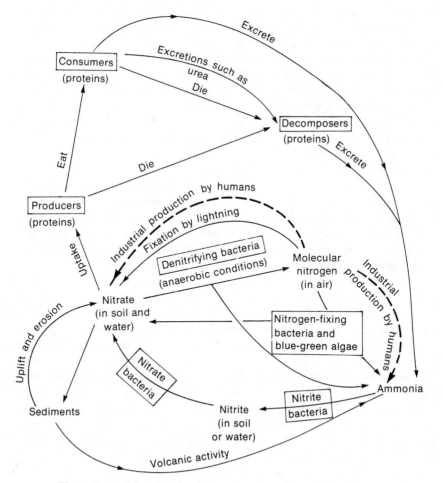

Figure 4–11. Main features of the nitrogen cycle in the biosphere.

poisonous and many larger organisms convert ammonia to some other, less toxic compound which can be stored for a time. In mammals such as man ammonia is converted to urea. Urea and similar compounds contain energy useful to certain decomposers, which break them back down to ammonia.

For the nitrogen to become available to green plants again the ammonia must be converted to nitrate. Conversion to nitrate occurs through the activity of two groups of bacteria, the *nitrite bacteria* which convert ammonia to nitrite and the *nitrate bacteria* which convert nitrite to nitrate.

This completes the basic circular system, but there are subsystems involving other microorganisms. *Denitrifying bacteria* convert nitrates to the molecular nitrogen that occurs in the atmosphere. *Nitrogen-fixing bacteria* (and blue-green algae) convert molecular nitrogen to ammonia or nitrates. The most familiar nitrogen fixers are the bacteria that live in nodules on the roots of legumes such as peas and clover but there are many types of free-living nitrogen fixers.

From the nitrate cycle it is clear that biogeochemical cycling is essential for the functioning of the ecosystem. Nitrate essential for plant growth becomes available primarily through the activities of organisms participating in the nitrogen cycle, which points up the complex interdependence of organisms in the ecosystem. We say that green plants are autotrophs, able to make their own food from simple inorganic materials. In one sense this is true but the nitrate necessary for green plants is the result of activities by other organisms.

The interdependence exists, of course, for many other elements. A world without decay organisms is scarcely conceivable but if you can imagine it—the ground beginning to pile up with dead leaves, trunks, excreta, and carcasses—you can see that nutrients would be tied up in these dead materials. Their release, through purely physical and chemical decomposition processes, would be so slow that primary production would have to decrease. The carbon cycle illustrates that no one area or ecosystem is independent of another. The carbon dioxide content of the air over a given forest involves not only the events in that forest but also many other things happening elsewhere in the biosphere—such as burning of fossil fuels and the degree to which the carbon dioxide is removed by the ocean.

Watershed Studies. Carbon and nitrogen are both elements which occur in a gaseous form that constitutes a fair part of the atmosphere. In these cases it is easy to see how movement of the elements from place to place over the globe can occur, but what of other elements such as phosphorus, calcium, and sulfur which are not present in significant amounts in the atmosphere? These are released from rocks by weathering and become available for plant use. They tend to be leached from soil and carried into streams and eventually to the ocean, where they may be deposited as sediments. Will these materials eventually all be lost from the soil to the oceans? In the geological long term it is expected that there will be uplift of some areas now covered by ocean (and subsidence of some areas that are now dry land) as in the past. Soil will then be made from the material accumulated as sediments and the minerals will again become available for use by organisms.

Recent studies of nutrient relations within single watersheds have led to a clearer picture of the cycling of these elements, as well as those with a gaseous form. The nutrients lost from natural ecosystems may be very small. From a hectare (about 2½ acres) of mature northern hardwood forest around Hubbard Brook in New Hampshire about 12 kilograms (26 pounds) of calcium, 16 kilograms (36 pounds) of sulfur, and less than 2 kilograms (4 pounds) of potassium were lost over the period of a year in stream water. Another finding is that considerable quantities of nutrients are returned to ecosystems from the air, mostly through rainfall. The nutrients get into the air mainly from the oceans, through such processes as wave action and spray. Also, for some materials such as sulfur and various trace elements, man's activities, especially burning fossil fuels, have become important. At Hubbard Brook inputs for calcium, sulfur, and potassium were large enough that the net losses were only 9, 3, and 0.5 kilogram per hectare, respectively. Presumably losses such as these can be made up through weathering in the soil. A diagram illustrating the cycling of calcium is given in Figure 4–12.

Losses from ecosystems such as crop fields where there is little vegetation, litter, or humus may be large, and the same is true for forest areas that are clearcut or burnt, although in the forests much of the output is from nutrients that had been stored in the vegetation.

Unlike crop fields mature ecosystems are tight; they do not waste nutrients. However, it should not be concluded from this that the streams flowing out of a watershed covered by mature forest will have no nutrients or even that the nutrients will be lower than in the stages just preceding. Logically, this is not possible, as W. B. Leak of the U.S. Forest Service has observed. Fully mature forests are at a steady state, not adding new biomass or accumulating litter. If there is any input of nutrients by rainfall or by weathering there is no place for them to be retained; consequently, an amount equal to the input must be carried away. Successional communities, on the other hand, which are accumulating biomass and litter will retain large amounts of the minerals they use in making protoplasm so that the output will be less than the input (atmospheric plus weathering). P. M. Vitousek and W. A. Reiners have put these ideas into a formal model of how watershed outputs should be related to the maturity of a community (Fig. 4–13). In early development there should be essentially no output of limiting nutrients such as nitrogen and phosphorus. Elements that are not essential for plant growth may be simply passed through. Some, however, could be stored on clay or humus. In the mature

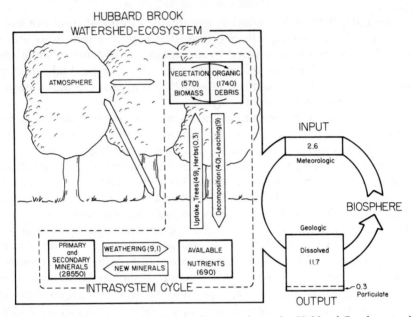

Figure 4–12. Major features of the calcium cycle in the Hubbard Brook watershed ecosystems. Figures are in kilogram/hectare. Vegetation contains 570 kg of calcium and an additional 1740 kg calcium are in litter, humus, etc. At any one time, about 690 kg are available in the soil for plant growth. Rain water supplies 2.6 kg per hectare per year and 12 kg is taken away via Hubbard Brook. Consequently, only 9.1 kg needs to be added to the soil by weathering to keep the system balanced. (From G. E. Likens and F. H. Bormann, "Nutrient cycling in ecosystems," in *Ecosystem Structure and Function*, J. A. Wiens, ed., Corvallis, Oregon State University Press, 1972, p. 53.)

ecosystem, output should equal input for all elements. If an ecosystem is heavily disturbed, as by clearcutting or fire (the broken line in the figure), output of the limiting nutrients will rise sharply as nutrients formerly stored in biomass are carried away.

The return of at least one essential element, phosphorus, from the sea seems to be a very slow process. Of course, phosphorus is very scarce in both earth and sea anyhow. Perhaps it will be found that it is somehow returned in small quantities in the same way as sulfur. But at present, it seems as though about the only way that phosphorus returns from the sea is through the landward migration of animals who spend time in the ocean, such as salmon or gulls.

The Hydrological Cycle. The seaward movement of materials such as phosphorus and calcium occurs in connection with one of the most basic of material cycles on the earth, the *hydrological cycle* (Fig. 4–14). This is the movement of water between ocean, earth, and atmosphere. The basic pattern of circulation is that precipitation falls on the earth, mainly as rain, although snow is important in some areas. Some of this water is returned to the atmosphere through evaporation and transpira-

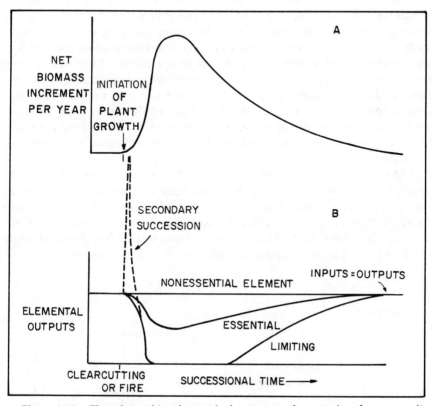

Figure 4–13. The relationship of watershed outputs to the maturity of a community. (From P. M. Vitousek and W. A. Reiners, "Ecosystem succession and nutrient retention: a hypothesis," *BioScience*, 25:377, 1975.)

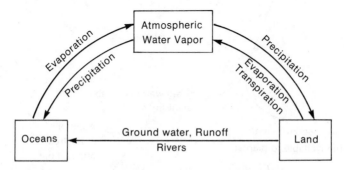

Figure 4–14. Main features of the earth's hydrological cycle.

tion (Fig. 4–15). For the land areas as a whole less water is returned in this way than falls as rain; the excess soaks into the ground or runs off. Whatever is not used somewhere along the way eventually makes its way through streams or ground water to the ocean (carrying with it elements such as phosphorus, sulfur, and calcium). Something more than 98% of the earth's available water is contained in the oceans. Evaporation from the ocean's surface, as well as evaporation and transpiration from the land, adds water vapor to the air. The water vapor in the air at any one time amounts to only about 0.001% of the earth's water, equal to about an inch of rain over the whole earth's surface if it were all precipitated at once. It is from the water vapor in the air, nevertheless, that come the rains that make plant growth possible.

Much of the rain falling on a vegetated area is intercepted by the vegetation and reevaporated without reaching the ground. Depending on the rain, the temperature, and the vegetation this may vary from not much to 100%. Some of the rain which reaches the ground runs off on the surface. Generally, surface runoff on an undisturbed soil does not begin until the soil has absorbed a fair amount of water. Some of the water

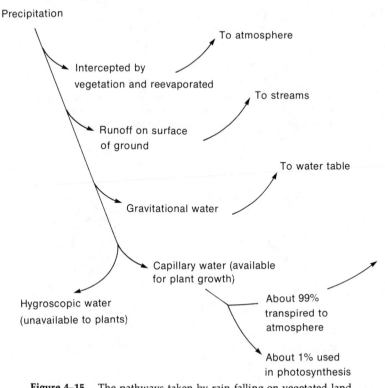

Figure 4–15. The pathways taken by rain falling on vegetated land.

absorbed by the soil (*gravitational water*) will move on downward through the ground to the water table. It is temporarily available to plants as it goes by. The water remaining after the rain is over and the gravitational water has drained away is that which is generally available for plant growth. This is *capillary water*, held in the small spaces between the soil particles. (There is also an additional small fraction, *hygroscopic water*, held in a tight film around the soil particles which the plant roots cannot remove.)

In discussing carbon and nitrogen we traced parts of the pathways of these elements within the organisms. For water, however, we can virtually ignore such things. This is not because water is unimportant to the organisms but because more than 99% of the water taken in by the roots of a plant is transpired again as water. Less than 1% becomes involved in the complicated biochemical pathways of photosynthesis. The situation is comparable in animals; although water serves a variety of functions, it leaves the body mostly as water and in about the quantities in which it was taken in.

In terms of the global hydrological cycle, organisms do not play roles as fundamental as in the carbon, oxygen, or nitrogen cycles. Their role seems mainly to be in modifying rates of movement through such activities as slowing runoff and increasing the water storage capacity of areas.

THE WORLD'S GREATEST REACTION

It has taken the effects of pollution to make most of us realize the power man has to alter the earth, its oceans, and its atmosphere. But it is only this realization and not the effects that are new. The earth and the life on it have evolved together, each affecting the other. The atmosphere that man has known since he evolved is a biological product.

Current theory suggests that the earth was formed somewhat less than 5 billion years ago. The first atmosphere, consisting of elements that are gases at moderate temperatures, was evidently lost, probably by escaping from the earth's gravitational field. This conclusion was reached because the earth, in comparison with the rest of the solar system, is deficient in most of the light elements such as hydrogen, helium, neon, and argon.

The second atmosphere of the earth is believed to have been produced by the release of gases from the earth's interior (called outgassing by geophysicists), mainly by volcanic activity. This process is still going on. The secondary atmosphere presumably consisted of compounds that are found in volcanoes today, along with compounds formed from them. Water, carbon monoxide, hydrogen, nitrogen, carbon dioxide, sulfur dioxide, and hydrochloric acid were probably present. So also may have been methane and ammonia. This is debated by students of the earth's early history but there is general agreement that the atmosphere was a *reducing* atmosphere—that is, there was no free oxygen but instead an excess of hydrogen.

With the accumulation of water oceans would begin to form, containing dissolved gases and salts. It is generally believed that life originated in the primitive oceans. It may have occurred in a fashion similar to that originally proposed by A. I. Oparin and J. B. S. Haldane, although the details of their reconstruction (as well as more recent modifications) may well be incorrect.

The details of the origin of life are not of major concern at this point. The model basically supposes chance collisions of inorganic molecules occurring in the presence of a suitable energy source.* Some of these collisions would produce more complicated molecules. The probability of producing a compound as complicated as an amino acid in this way in one swoop is low but the probability of producing less complicated molecules is higher; once these are formed the probability of their coming together to form still more complicated molecules is reasonably high. Given a glass flask containing water and a methane, ammonia, and hydrogen atmosphere, and with energy supplied in the form of heat and an electric spark, you might suppose that it would take years, or even forever, to produce measurable quantities of amino acids, fatty acids, and sugars. In fact, in an experiment performed by S. L. Miller in 1953, it took a week.

Since then many other similar experiments have been performed under various conditions including lower temperatures, ultraviolet radiation as an energy source, and many combinations of gases. In the presence of a reducing atmosphere the production of many kinds of organic compounds occurs rapidly.

The step from organic compounds to something that could be called an organism is a long one, but there was a long time available for the step (or rather, the very many small steps) to have occurred. The first organisms are believed to have been some sort of single-celled heterotrophs. Their food, which they broke down through anaerobic respiration, consisted of the various organic compounds produced in the way already described, which had made the ocean a dilute soup by this time.

With the increase in numbers of these primitive heterotrophs competition would develop. The competition would be for food—for the rather complex organic molecules that had been produced in the millions of years of chemical evolution. A change allowing the use of some simpler compound by one of the primitive heterotrophs would give it an advantage, since it could then exploit a food source not available to other

---

*This might have been ultraviolet radiation, which would have been more intense at this period. Today most of the ultraviolet radiation is screened out before reaching the earth's surface by an ozone layer in the upper atmosphere. But the ozone layer is produced by the action of sunlight on oxygen.

primitive organisms. As these simpler materials became depleted further changes allowing the use of still simpler materials would be favored. The eventual result would be the development of organisms able to manufacture all their required organic materials from inorganic materials. In this sequence of events lies the origin of autotrophs.

With the evolution of photosynthetic autotrophs there was a steady source of free oxygen for the first time on earth. The oxygen accumulated slowly. Some of it oxidized metals such as iron; the earliest "red beds" indicating deposition of iron-containing sediments with appreciable oxygen in the water are dated at about 2 billion years ago. As oxygen accumulated in the air an ozone layer would be produced at high altitidues, decreasing the amount of ultraviolet radiation reaching the earth's surface and making the upper levels of the oceans (and eventually the land) habitable. The accumulation of oxygen in the atmosphere made a conversion from anaerobic to aerobic respiration possible. Aerobic respiration yields 19 times as much energy per molecule of sugar broken down as does anaerobic respiration, so that those organisms which developed mutations allowing aerobic respiration began to win out over their anaerobic relatives. The evolution of plants and animals as they are known today could then begin.

The point is that our present atmosphere is of biological origin. The development of photosynthesis, with the liberation of oxygen, marked the beginning of the greatest reaction in the earth's history, a revolution that made possible the development of advanced life.

ORGANIZATION OF COMMUNITIES
AND ECOSYSTEMS

DOMINANCE

Organisms that exert more control than others on the character of the community of which they are a part are called *dominants*. The control can be exerted through reaction or coaction. Trees are obvious dominants of forests that create, mainly through reaction, a microclimate suitable for a certain group of organisms and, accordingly, set the character of the community. Cattle are dominants of an overgrazed grassland; certain kinds of plants that are preferred by the animals or that cannot withstand being trampled have disappeared and been replaced by plants that are spiny, ill-tasting, or tough. Here the control is largely through coactions.

Dominance is not all or nothing, since virtually every organism of a community plays some role in determining its nature down to the most insignificant microbe. In fact, some insignificant microbe of the forest may be just as important as the trees if the microbe is, for instance, essential for leaf decay.

CHEMICAL ECOLOGY

It seems apparent that a considerable amount of the organization of ecosystems is based on chemical interactions. The term *chemical ecology* refers to the study of the production and uptake or reception by organisms of chemical compounds having effects on other organisms (either intra- or interspecifically). Chemical ecology is not basically different from any other kind of ecology but it is worth a separate discussion at this time, because only recently has enough attention been given to chemical inter- actions for their importance to be appreciated.

Of course, when a lion eats an antelope's leg, it is eating chemicals, and when we breathe air we are breathing oxygen and carbon dioxide produced by other organisms. Generally, chemical ecology is not used to include simple relationships of feeding and the movement of inorganic nutrients. R. H. Whittaker and P. P. Feeny have set up a classification of interorganismic chemical effects that is used here (slightly modified). They have separated such effects into effects between different species (which they call *allelochemic* effects) and effects between individuals of the same species. The second category includes the *pheromones* that serve as chemical messengers between members of a species. These may be compounds wafted off in the air that attract male moths to female moths, or compounds in urine that mark the territorial boundaries of a wolf. In the form of royal jelly, they may determine whether a honeybee larva becomes a worker or a queen.

The allelochemicals have a great variety of effects. There are *repellents* such as the distasteful and poisonous compounds in buttercup leaves or mustard and the noxious-smelling butyl mercaptan sprayed by the skunk. There are other kinds of *escape substances*: some insects that live supported on the surface film of water release detergent-like substances to reduce the surface tension behind them, so that they are whisked rapidly forward and away from a predator.

There is the important *suppressant* group. Included are the materials produced by chaparral shrubs that inhibit herbaceous growth around them. In terminology used for competition this is an example of interfer- ence. Substances released by certain planktonic plants that slow the feed- ing rate of some herbivorous zooplankton are also suppressants. Included also are the antibiotics, produced mainly by various kinds of fungi, that man has used so successfully for his own benefit. Presumably in the soil where these organisms live the antibiotic substance functions to inhibit the growth of competitors, another example of interference. In some cases the suppressant material eventually has inhibitory effects on the species producing it; it is then called an *autotoxin*.

Attractants may serve an aggressive function as in the case of attrac- tants which lure insects to carnivorous plants, or a nonaggressive one, as in flower scents which bring insects to flowers. The attractant scent may benefit both the flower which is pollinated and the insect which obtains pollen or nectar. In this case the whole interaction would be a case of mutualism. But when mosquitoes use the lactic acid on our skin as a signal to bite the attractant is only to their advantage.

The same sort of coevolutionary dance has been performed at the chemical level as at other levels. Plants of the mustard family produce mustard oils that are irritants to most animals, but some insects have evolved the ability to feed on the mustard plants and some even use the mustard oils as cues for locating the plants. Furthermore, a wasp that parasitizes an aphid that feeds on the mustards uses the scent of mustard oil to find the aphids.

In a marvelously complex system the female polyphemus moth makes use of a compound from red oak leaves (received through her antennae, not by eating the leaves) as a signal that it is an appropriate time to mate. She then releases a pheromone received by the male polyphemus. The result is that larvae are produced at a time when foliage is suitable for them to feed on.

Probably chemical interactions are especially numerous in soil and water, but the number and variety of interactions in terrestrial ecosystems is also quite impressive. Because humans are so large and depend so strongly on sight we have very little appreciation of the environment of plant and animal chemicals around us. We move through a world of chemical scent trails of ants, moth sexual attractants, poisonous plants, and the varied scents of dogs, cats, rabbits (and people), and we are almost totally unaware. Only when we smell flowers, use coffee, tobacco, or cinnamon (in doses that would kill a grazing insect), or perhaps take our dog to a new neighborhood and watch it explore with its nose instead of its eyes, does this world begin to infringe on our consciousness.

SPATIAL STRUCTURE

Ecosystems generally have a noticeable vertical structure; that is, they have *layers* or *strata*. Many ecosystems show two broad trophic strata, an upper autotrophic and a lower heterotrophic. In a forest most food-making activities occur in the upper levels where the leaves are concentrated, and most consumption and decomposition occur on or beneath the forest floor. In a lake also food production occurs mainly in the upper part of the water where light can penetrate. Not many food-making activities take place at the bottom, but here are large numbers of animals and bacteria that live on the dead material that settles to the bottom. Of course, the layers are not separate and distinct; herbs make food on the forest floor and caterpillars and birds forage in the canopy.

Within or overlapping these trophic strata other structural strata may be distinguished. In many forests it is convenient to recognize the following layers, going downward: *trees, shrubs, herbs, floor,* and *subterranean.* Other communities may lack some of these or may have other strata. Three subterranean layers can be recognized in prairie, based on the depth reached by different kinds of plant roots (Table 4–4). Tropical rain forests may have three tree strata, but herb and shrub layers are poorly developed.

Organisms may divide the habitat up in a more complex manner than by just occurring in layers. Three different species of nuthatches live together in the ponderosa pine forests of Colorado in the winter. One of these is the familiar white-breated nuthatch, generally seen scrambling around a tree trunk probing for food in crevices in the bark. This is the way it also behaves in the Colorado pine forests. A second slightly smaller species, the red-breasted nuthatch, does most of its foraging on large branches of the trees and a third, still smaller species, the pigmy nuthatch, obtains most of its food from small branches and clusters of pine needles. Such division of resources presumably reduces interspecific competition (Chap. 3), and is part of the differences between the ecological niches of the three species (see pp. 158-162).

DAY-NIGHT CHANGE

Most organisms have **daily cycles**; organisms that are active in the daylight hours and asleep or inactive at night are *diurnal*, and those that are active at night and inactive by day are *nocturnal*. Some organisms are most active at dawn or dusk or both; these are said to be *crepuscular*.

These broad patterns of activity in the sense of being awake and running about are not the only 24-hour cycles that may be occurring in an organism, however. There may also be daily patterns of physiological activity such as the production of new cells or the secretion of a particular enzyme. Many of these cycles, whether of wakefulness, locomotion, or

Table 4–4 UNDERGROUND STRATIFICATION IN PRAIRIE*
The University of Nebraska student of the prairie J. E. Weaver specialized in the relationships of plants to the soil. He recognized three underground layers based on the depths to which roots penetrated.

Layer	Depth (m)	Per cent of species	Examples
Shallow	0.6	14	Blue grama, June grass
Medium	0.6–1.5	21	Needle grass, buffalo grass, many flowered aster
Deep	1.5–6	65	Big bluestem, slough grass, compass plant

*From Weaver J. E., and Clements, F. E., *Plant Ecology,* 2nd ed., New York, McGraw-Hill, 1938, p. 315, and other publications by Weaver.

more subtle activities, will persist even during an experiment when the organism is placed in an environment with no alternation of light and dark, and with all other controllable factors such as temperature and humidity held constant. Daily cycles that continue under these constant conditions are called *circadian rhythms*. The persistence of such rhythms implies a time-keeping ability on the part of organisms, who are said to have a "biological clock." Without the cue of sunrise every day the biological clock of organisms may run somewhat fast or slow, so that individuals kept under constant conditions tend to have cycles that are a little longer or shorter than 24 hours.

Humans are strongly diurnal creatures. Even those "night people" who sleep until noon and work or watch late movies until the early hours do not go walking through the forest at night; humans are just not built for it. Consequently few of us are aware of the day-night pattern of ecosystem organization.

Physical conditions at night are different—it is not just darker but cooler and more humid. The animals that are encountered at night are different. Except for owls and whip-poor-wills (and their relatives) birds are asleep. On the other hand mammals are abroad. According to the calculations of the Northwestern University ecologist O. Park, one of the few students of nocturnalism, 60 or 70% of forest mammals are nocturnal. And the types of invertebrates that are active are different. The moist-skinned earthworms and slugs are out, lightning-bugs may be flashing, if it is early summer, and crickets and katydids may be calling (by rubbing their forewings together) if it is late summer.

Plants are not photosynthesizing at night but they are respiring and translocating the food produced in the leaves during the day downward. Still other events are occurring; some plants show "sleep movements" in which the leaves fold up at night. Although most plants open their flowers either diurnally or throughout the entire 24-hour day some species only open their flowers at night. Most of these flowers are moth-pollinated and are white, which makes them more conspicuous in dim light.

In any case many nocturnal animals lack color vision. Nocturnal species of birds and mammals tend to have more rods and fewer cones in the retina of their eye as compared to diurnal species. Rods are stimulated by lower intensities of light so vision becomes more efficient in near darkness. There are other features that nocturnal organisms have that seem to be adaptations to low light levels. If a light is shined on a raccoon or a whip-poor-will its eyes will glow. Eye shine is characteristic of many nocturnal species but few diurnal ones. It is light reflected from a layer in the back of the eye (the *tapeta lucidum*) and probably aids vision by, in effect, running the light through the retina twice, rather than by just absorbing it at the back of the retina.

There is no single answer as to why some organisms or some activities are nocturnal or diurnal. For slugs and salamanders it may be no more than a matter of high humidities at night being more favorable. For many organisms evading the many diurnal predators may be important, or avoiding a diurnal competitor might be the underlying evolutionary advantage. This would most likely be of importance between organisms in which actual interference in the use of resources was involved.

SEASONAL CHANGE

Most ecosystems show **seasonal changes** in structure, appearance, and function which are dependent on seasonal changes in the physical environment, especially temperature, precipitation, and photoperiod. As a community the tropical rain forest shows little in the way of seasonal changes. "All year round," wrote one student of tropical forests, P. W. Richards, "the foliage is the same sombre green and in every month some species are in flower."* But even in tropical rain forest the individual species tend to be periodic in their activites, and for many activities the period may be approximately annual.

In the Deciduous Forest

In contrast to tropical rain forest, seasonal changes in the deciduous forest are probably as marked as anywhere on the earth. Six recognizable seasons, each blending into the next, and some of their major events are described in this section.

Early Spring (Prevernal Aspect). The buds of trees begin to enlarge; maple sap is running. The earliest spring flowers, such as harbinger-of-spring and skunk cabbage, bloom (Table 4-5). The earliest spring migrants, such as red-winged blackbirds and eastern phoebes, return. On sunny mornings, cardinals and tufted titmice sing. The big owls, barred and great horned, are nesting. Spring peepers begin to call in the marshes.

Late Spring (Vernal Aspect). The later spring flowers, such as white trillium and May apple, bloom, as does flowering dogwood. Trees leaf out, maples early, ashes and oaks late. This is the peak period of bird migration but many birds are already beginning to nest. There is a local migration of insects from overwintering sites in the forest floor to summer habitats. The spring overturn occurs in lakes and there are fish runs upstream for spawning.

Summer (Aestival Aspect). This is the height of bird nesting as most species finish incubation and begin feeding young. Consumption of tree leaves by caterpillars (and caterpillars by birds) peaks. Flies and mosquitoes are numerous. Many forest flowers, the spring ephemerals such as

---

*In *The Tropical Rain Forest. An Ecological Study*, New York, Cambridge University Press, 1952, p. 191.

Table 4-5 WHEN DO THE SPRING FLOWERS BLOOM?*

Long records of the dates of seasonal events are scarce. Aldo Leopold kept about ten years of such records for Dane County (where the University of Wisconsin is located) and Sauk County (the Sand County of *Sand County Almanac*). Probably the longest record was kept by Thomas Mikesell. Around the small town of Wauseon in the northwestern corner of Ohio, he kept track of the dates of such events as flowering, leafing, and fruit ripening for many species of plants from 1883 to 1912. His average dates for the first blossom of some common spring flowers are given below.

Species	Average Date of First Blossom
Hepatica	April 13
Spring beauty	April 17
Bloodroot	April 18
Yellow trout lily	April 23
Pepper root	April 23
Early blue violet	April 25
Dutchman's breeches	April 28
Yellow violet	May 2
Wild blue phlox	May 2
Wake-robin	May 5
Jack-in-the-pulpit	May 14
May apple	May 16

*From Smith, J. W. "Phenological dates and meteorological data recorded by Thomas Mikesell at Wauseon, Ohio," U.S. Department of Agriculture, *Monthly Weather Rev. Suppl.*, 2:23-93, 1915.

Dutchman's breeches, trout lily, and spring beauty, die back entirely above ground and rest as corms or tubers until the next spring.

Late Summer (Serotinal Aspect). The roadsides are colorful with many open country plants blooming—black-eyed Susan, chicory, and turk's-cap lily. Buckeye leaves die and fall. A few birds such as the goldfinch and the cedar waxwing are just beginning to nest, but family groups of many species are wandering about, and there is little territoriality and the singing that goes with it. Young spiders and toads are abundant. Cicadas emerge from the ground and become vocal.

Fall (Autumnal Aspect). The late open country flowers such as blazing stars and asters bloom; this is also a time of flowering in the marsh where New England aster, fringed gentian, and grass-of-Parnassus may be found in bloom. In the forest nuts such as acorns and beechnuts ripen. Many of the herbs have ripe fruits of the beggar's-tick or sticktight variety. Tree leaves turn color and fall, with willows, oaks, and tamaracks among the last. Fall bird migration is at its heaviest. There is a migration of forest insects down into the forest litter and soil, and a migration of open country insects into the litter and soil of the forest edge.

Winter (Hiemal Aspect). Most plants are dormant. Only winter and permanent resident birds are around; most of these gather in flocks. Also still active are most aquatic animals and some mammals such as mice, shrews, and foxes. Otherwise most animals are hibernating, such as many invertebrates and the woodchuck, or sleeping, such as the chipmunk.

In Other Ecosystems

In many tropic and subtropical ecosystems wet and dry conditions are more important in seasonality than warm and cold. This is true for the deciduous forests of the tropics, such as monsoon forests, and for many deserts (Table 4–6). In the more extreme deserts some rainy season events may not recur every year. In some years, possibly several in a row, not enough rain will fall to allow the growth of certain of the plants or to bring some animals out of the rest period (*aestivation*) in which they endure the drought.

Aquatic communities also show seasonal change. Even hot springs, which have constant temperature and salt concentration, show changes related to such factors as seasonal differences in sunlight. In the far north light may be so dim and brief that algae die back because they use more food in respiration than is produced by photosynthesis. Farther south, the algae may not change much in biomass but may increase their chlo-

Table 4–6 SEASONAL CHANGE IN THE SONORA DESERT

In contrast to the six seasons in temperate deciduous forest, the Sonora desert in the vicinity of Tucson has four seasons.

Season	Weather	Biological Events
Winter wet (December–March)	Widespread rains totaling 2–3 inches; cool weather, last frost mid-March.	Vegetation shows signs of activity in January. Many plants, both winter perennials and winter annuals, bloom in February and have mature fruits in March and April. Bird nesting begins.
Dry foresummer (April–June)	Total precipitation about 1 inch; high temperatures.	Mesquite leafs out; however, many winter perennials lose their leaves during the course of the season and do not bear leaves again until the following January. Succulents — cactuses (including saguaro), yuccas, agaves—bloom in profusion. A peak of bird nesting, including most of the desert species. Also heavy migration of birds which nest further north.
Humid mid-summer (July–September)	Greatest precipitation of year in July and August coming, however, in local heavy thunderstorms; temperatures very high—July average over 85°F (about 30°C).	Desert transformed by growth and flowering of vast number of summer annuals. Grasses and some succulents such as the great barrel cactus flower. Spadefoot toads breed. A second peak of bird nesting, including the rufous-winged sparrow and blue grosbeak. Heavy bird migration continuing into November.
Dry aftersummer (October–November)	Little rain; warm to cool. First frost mid-November.	The grasses and some other plants flowering in the humid midsummer ripen seeds. A nearly complete cessation of vegetative activity.

rophyll content, evidently compensating for the decreased light. One of the best known sequences of seasonal change in an aquatic situation involves the thermal stratification of lakes (see the section on aquatic habitats in Chap. 5).

Phenology

The scientific study of seasonal change is called *phenology*, perhaps the earliest branch of ecology. A Chinese calendar from the Hoang Ho Valley giving the dates of biological events is known from 700 B.C. Primitive humans often planted crops according to the flowering times of wild plants. But despite its ancient origins phenology has not received much attention in ecology. One of the few efforts was made about 50 years ago by the entomologist A. D. Hopkins. He gathered together a vast amount of material on the timing of seasonal events to answer the question: How fast does the spring move north? Biologists indulge less in giving formal names to generalizations than chemists, but you may refer to *Hopkins' Bioclimatic Law* if you wish. The "law" is not perfect, since early spring is different from late spring, the North is different from the South, and each species is different from every other. Nevertheless, it is a widely useful rule of thumb.

Hopkins stated that in the spring events occur about 4 days later for every degree of latitude northward, meaning that spring moves northward at about the rate of 17 or 18 miles a day. It means that, compared to Minneapolis-St. Paul, a given event of the spring occurs 2 weeks earlier in Des Moines, 26 days earlier in St. Louis, and almost a month and half earlier in Little Rock. Altitude also has it effect, with events occurring a day later for every 100 feet of elevation. In mountainous regions you can go from spring back to winter in a morning's hike from a valley up one of the peaks. Longitude is also significant, with events occurring one day later for every 1¼° of longitude westward. Spring comes earlier to Philadelphia than it does to Columbus, Ohio, even though both are on the fortieth parallel of latitude.

All of this says nothing about why the phoebe migrates or the sugar maple leafs out when they do. That is another question or, rather, another two questions involving the organism and the information it receives from the environment and the evolutionary reasons for the scheduling of events (proximate and ultimate factors discussed in Chap. 2).

There are some obvious practical functions of phenological data that have been appreciated more by the general public than by scientists, at least in recent years. Planting corn when oak leaves are the size of squirrel's ears is not a bad way to avoid frost while getting the crop started early enough for the ears to ripen. In prepesticide days, agricultural experiment stations published calendars of the date of emergence of the Hessian fly. If winter wheat plants are present the flies lay eggs on them and the larvae, feeding on the sap of the plant, greatly reduce the wheat yield. However, the adult flies do not live long and by simply planting wheat after the emergence of the adult flies, damage to the wheat crop is prevented. The potential for this use of easily observed events as *ecological indicators* of other, economically more important, events has scarcely been tapped.

ECOLOGICAL NICHE

The term *niche* was first used at about the same time but in slightly different ways by the English animal ecologist Charles Elton and the University of California naturalist Joseph Grinnell. Attempts in the 1960's to produce a rigorous definition of the concept (pp. 160-162) have not lived up to their promise but the term seems valuable as a kind of shorthand expression indicating the entire role played by a species in an ecosystem. "Tree trunks" might specify in a general way the foraging dimension of a bird's niche, "insect eggs" the food dimension, and "natural cavities" the nesting dimension (Table 4–7). In discussing niches we are viewing the community as a unit made up of the niches of the various species composing it.

In comparing different but similar communities we are sometimes struck by the similarities of niches. In deserts, for example, there seems to be a "succulent" niche filled in North America by cactuses and in Africa by euphorbias. According to E. P. Odum the "bottom-dwelling carnivore" niche is filled in different coastal areas around North and Central America by the spiny lobster (tropical), king crab (upper West Coast), stone crab (Gulf Coast), and lobster (upper East Coast). Species filling similar niches in different regions are called *ecological equivalents* or *ecological counterparts*. The physical resemblance between the equivalent species may be only slight (as between kangaroos and bison, considered to be counterpart grazers in North America and Australia), but can at times be remarkably close. The yellow-throated longclaw, a bird of African grassland, bears a very striking resemblance to our meadowlark but is not closely related to it.

It would, however, be a mistake to think that every deciduous forest or desert or tropical rain forest has the same niches. The occasional examples of spectacularly similar counterpart species are at least balanced by cases of similar communities in which resources are divided up very differently.

THE NICHE AS A HYPERVOLUME

In this section is described the concept of the *ecological niche* as put forth by G. E. Hutchinson of Yale University, possibly the most influential

Table 4–7 BIRD NICHES IN CLIMAX MESIC DECIDUOUS FOREST

Two dozen niches are listed, mostly based on layer and type or size of food. "Large" animals eat, on the average, larger food items than "small" but also tend to be more versatile, taking some smaller food items.

Niche	Species
Large invertebrates on forest floor	Wood thrush
Small invertebrates on forest floor	Ovenbird (north), Kentucky warbler (south)
Understory gleaner	Hooded warbler

There seem to be more niches at the floor-understory level in southern forests. In addition to the Kentucky warbler and the hooded warbler (itself much more common in the South), the worm-eating warbler and Carolina wren are floor-understory species there.

Generalized forager on shrub and low tree invertebrates	Black-capped chickadee (north), Carolina chickadee (south)
Generalized forager on medium and high invertebrates	Tufted titmouse
Large canopy gleaner	Scarlet tanager
Medium canopy gleaner, leaves	Red-eyed vireo
Medium canopy gleaner, twigs	Yellow-throated vireo
Small canopy gleaner	Cerulean warbler
Very large woodpecker	Pileated woodpecker
Large woodpecker	Red-bellied woodpecker
Medium woodpecker	Hairy woodpecker
Small woodpecker	Downy woodpecker
Tree-trunk prober	White-breasted nuthatch
Small, low static flycatcher	Acadian flycatcher
Medium, high static flycatcher	Eastern wood pewee
Large, static flycatcher	Great crested flycatcher
Moving flycatcher, low	American redstart
Moving flycatcher, high	Blue-gray gnatcatcher
Large seeds and fruits	Cardinal
Small seeds and fruits	Indigo bunting

Both the cardinal and the indigo bunting tend to be species of openings in the forest. Because openings are natural, recurring from such causes as wind-throw, both species are logically considered forest species. The scarcity of seed eaters may be related both to relatively low production of seeds and to high populations of seed-eating mammals such as deer mice and chipmunks.

Large omnivore	Blue jay
Very large omnivore	Common crow
Large owl	Barred owl (wetter forests)
	Great horned owl (drier forests)
Nectar feeder	Female ruby-throated hummingbird

Some other possible niches that seem to support species only occasionally or at low densities are large browser (ruffed grouse), small owl (saw-whet owl), large hawk (red-tailed or red-shouldered hawk), and medium hawk (Cooper's hawk).

ecologist of the mid-twentieth century. The general idea of the concept is given by the title of one of his essays, "The niche: an abstractly inhabited hypervolume."* The intellectual atmosphere surrounding the development of the concept should be remembered—it was one in which the competitive exclusion principle was held as an important ecological rule. This concept of the niche was developed as a context for the competitive exclusion principle.

Suppose that we measure the range of some environmental variable (say, temperature) over which a particular species can live and reproduce (in effect, its range of tolerance) and we put this on a graph:

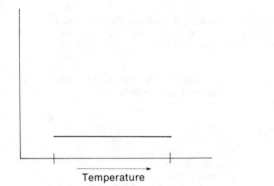

Temperature

Suppose that we then do the same for another environmental variable (humidity, perhaps) and put this on the second axis of the graph. The space that is enclosed is a rectangle that represents two of the dimensions of the ecological niche of the species:

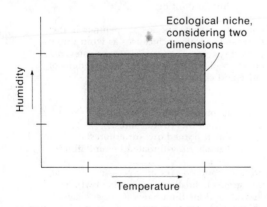

Ecological niche, considering two dimensions

Humidity

Temperature

*In *The Ecological Theater and the Evolutionary Play,* op. cit., pp. 26–78.

If we now erect a third axis for a third environmental variable (perhaps some nutrient) the space that is enclosed is now a volume in three dimensions:

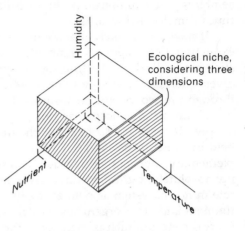

Ecological niche, considering three dimensions

If we now erect a fourth axis for a fourth environmental variable, the space enclosed is a hypervolume having four dimensions. Also, it gets hard to draw. We may include more and more dimensions, up to n, so that the niche as defined by Hutchinson has an infinite number of dimensions. This is the *fundamental niche* of the species. If the fundamental niches of two species overlap, then the two species are competing. Note, however, that an apparent overlap in two, three, four, or ten dimensions may not mean any actual overlap; adding one more dimension may move the hypervolumes far apart. (For an analogous situation, hold one hand a few inches in front of your eyes and one at arm's length. The two hands seem to overlap in two dimensions but you know that in three dimensions they are separated by a foot or so.) Two species, for example, do not overlap, or compete for food, even though they eat items of exactly the same size if they look for them in different places.

Hutchinson viewed the outcome of interspecific competition as either extinction or the development of differences allowing coexistence. Accordingly, he recognized a *realized niche* which was a hypervolume bounded by the actual ranges, as found in the field, for each variable. To the degree that a species was excluded by competition from situations it was potentially able to use, its realized niche would be smaller than its fundamental niche.

Most well integrated communities (such as a climax forest or a coral reef but possibly not a first-year weed field) would be made up of species with nonoverlapping niches. The defunct *"broken stick" model* of Robert MacArthur, Hutchinson's best-known student, dates from this period. In this the community is represented by a stick and the species are segments formed by breaking the stick into lengths corresponding to the abundance of each. It seems clear that during this period in the 1960's Hutchinson and MacArthur suspected that communities were made up of niches that fit together like the pieces of n-dimensional jigsaw puzzles (possibly resembling, in n dimensions, those little ball-shaped puzzles that sometimes come on key chains).

It now seems likely that some overlap can, and does, occur without extinction. Only if energy is included as a niche dimension is the broken-stick model likely to be true for all species of all communities. Since a given quantum of energy is respired once and only once in its passage through the ecosystem, niches must be nonoverlapping in this sense.

If the niche concept has not proved useful in itself, then why not simply abandon it? To begin with, the overall view of the community held by Hutchinson (and S. C. Kendeigh, among others) of largely complementary, noncompetitive use of resources seems basically, although not absolutely, right. Second, a given species does have a sort of abstract role in an ecosystem as defined by its relationships to the physical environment and other organisms. *Ecological niche* is a useful shorthand term for this concept. Such is the case in the interesting work on *niche breadth* (see p. 165) which grew out of Hutchinson's concept of the niche which does not actually depend on it.

COEVOLUTION

Evolution through natural selection is a process of populations; however, the process goes on in the context of the ecosystem, including the other species of the ecosystem. This idea is captured very neatly in the title of a book by G. E. Hutchinson, *The Ecological Theater and the Evolutionary Play.**

In some cases two or more species interact so closely that evolutionary changes in one tend to be followed by evolutionary changes in the other, so that they form an evolving system. This evolving together was termed *coevolution* by P. R. Raven and P. H. Ehrlich. Coevolution probably occurs among many pairs (or larger numbers) of species lnked by coactions. Prey, as we suggested earlier, under the selective pressure of their predators, may tend to evolve toward running faster or hiding better. Predators, under the selective pressure of starvation, may become more stealthy or develop keener eyes or noses.

Coevolution can be involved in mutualistic associations; in fact, no other explanation seems possible for obligatory mutualistic associations. Reproduction of the yucca plant is dependent on the yucca moth, its only pollinator, but reproduction of the yucca moth is dependent on the yucca plant, in whose flowers its eggs develop. The same is true of the fig and

---

*New Haven, Yale University Press, 1965.

the fig wasp. At some time the relationship between these plants and insects must have been looser; the tight interaction in which one cannot exist without the other must have developed stepwise in coevolution.

An especially well-studied example of a mutualistic interaction is the ant-acacia system. In Central America certain kinds of acacia trees have thorns with very large bases. Members of an ant colony live within these thorns, which they hollow out. There are related types of acacias without swollen thorns which do not harbor ant colonies and related types of ants which do not live on acacias. Both the acacia ants and the swollen-thorn acacias have traits missing in their relatives that would seem to be of no benefit to them except in the context of their association.

The ants live and raise their young in the enlarged thorns. The adult ants feed on the nectar produced by the acacia, which is produced in nectaries located on the leaves, rather than in the flowers. Such nectaries are present on nonswollen-thorn acacias but are smaller than those on swollen-thorn plants. The larvae of the ants are fed material from specially modified leaf tips called Beltian bodies, which do not occur on nonswollen-thorn species.

The advantages to the ants of this coaction are obvious, but what advantages does the plant derive? It is protected from most kinds of herbivores, whether insect or mammal. Insects that land on the plant are attacked by the ants and killed or driven off. Likewise, a mammal that brushes against the plant is bitten and stung. The ecologist D. H. Janzen, who worked out many of the details of this case of mutualism, showed that these activities by the ants do, in fact, benefit the plant. He prevented ants from colonizing new sprouts and found that these sprouts grew slower, produced fewer leaves, and had a lower rate of survival than similar sprouts which the ants had colonized.

Unlike related species which are active only in the daytime the acacia ants patrol 24 hours a day, thus protecting the plant from both diurnal and nocturnal herbivores. The plants may gain another benefit from the ants' activities. The ants not only attack animals but also maul any other plants that touch the acacia foliage or that grow up below the acacia. Acacia is intolerant to shade and susceptible to fire, which is frequent in the area it inhabits. The clearing activities of the ants may tend to keep the acacia plant from being shaded out and may keep the vicinity clear of flammable material.

The ant-acacia interaction just described is obligate; this particular ant and acacia do not occur separately. There are, however, species of acacias and related species of ants in which the relationship is not as close; they often occur together but also may be found not associated with one another. Coevolution seems to be the most likely explanation for the obligatory relationship. The association must have arisen step by step through small changes which had the effect of fitting the ants and the acacia into a closer, more efficient system.

Possibly one of the most important kinds of coevolution has been described earlier (Chap. 3) in connection with interspecific competition. When a species extends its range and enters new communities, it may (probably very rarely) find unexploited resources and be able to establish itself immediately. On the other hand, it may be eliminated immediately by competition with the already existing species. A third possibility is that evolutionary adjustments may occur in one or both species, allowing **coexistence.** Coexistence is achieved by dividing up the shared resources in some way, perhaps by food being divided up so that one species feeds on larger items and the other on smaller. Or the species might divide the community up vertically, one feeding in the shrubs and one in the trees. Organisms may also divide the resources of the ecosystem *temporally* rather than spatially. For example, in grassland two similar herbs might diverge in such a way that one has its main growth and flowers in spring and the other in summer.

As G. E. Hutchinson and S. C. Kendeigh have both ably pointed out, many aspects of community organization described in the preceding sections seem to be based on the avoidance or reduction of competition. Most communities seem to be the product of this and other sorts of co-evolution occuring repeatedly over the course of geological history. The deciduous forest of eastern North America, with its stratification and seasonal change, its plants adapted to shadier and sunnier, wetter and drier sites, and its mycorrhizae and root parasites, is one such case.

ECOLOGICAL DIVERSITY

SPECIES DIVERSITY

A diverse community is one having a large amount of variety. Although there are other aspects to diversity the simplest measure is the number of species present; a diverse community has more species than a less diverse one. This aspect of diversity is sometimes called *richness*.

In theoretical terms, one community may have more species than another for the following reasons. (1) One community may not have all the species that could live there; it is *unsaturated*. Not all the potential niches are filled. (2) One community may have a greater range of one or more resources. For some resource, then, one community may look like the first graph at the top of page 165 and another, with a greater range of some resource and more species, may look like the second graph.

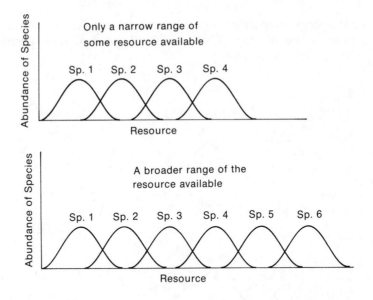

(3) Species in one community may be more specialized (that is, adapted to a narrower range of food, soil, or other environmental factors) than in another. In other words, they may have narrower niches. A community with narrower niches (and more species) compared with one with broader niches (and fewer species) is shown below. With narrower niches (and no increase in resources) each species will obviously be less abundant.

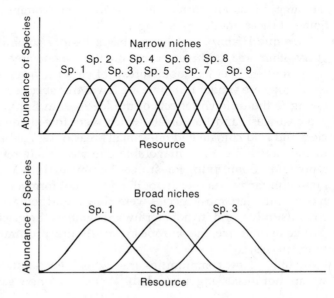

(4) Species may share resources to a greater degree; that is, there may be more niche overlap. With increased overlap, interspecific competition would tend to increase:

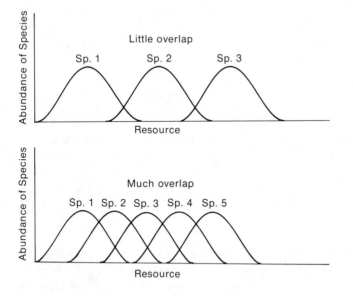

Factors Affecting Diversity

Many different influences have been suggested as affecting species diversity (in one or more of the ways just listed) and probably most of the suggestions are correct, at least for some community at some time. Some of these are the following.

Unique History. Each area has a history somewhat different from every other area, and past events may cause one area to be more diverse than another. The deciduous forests of eastern North America and western Europe are similar but there are fewer tree species in Europe, probably owing to the different effects of the Ice Age. In Europe mountain ranges run east-west, and the forests which were forced southward by the glaciers and cold climate had only small areas of refuge between the glaciers to the north and the unfavorable climate and terrain of the southern mountains. Some of the forest trees became extinct. In North America the mountain ranges run north-south and do not form a barrier to southward movement. Here most species were able to find refuges somewhere to the south, survive, and invade northward again as the glacier receded. In this particular instance the European forests are presumably not saturated with species.

Time. New communities arise when new habitats are created. Here we are not discussing a new habitat such as a new sandbar built by the

Mississippi River but a habitat of a type not previously available for organisms. Dry land would have been such a habitat when oxygen began to build up in the atmosphere and the continents became habitable. Presumably new communities have few species, consisting of those with traits that, even though evolved in some other community, allow them to survive in the new environment. With time more species would be added through various processes, including the adaptation of already existing species to the new habitat and the evolution of new species.

Extreme Habitats. Habitats that are harsh tend to have relatively few species. By harsh we mean habitats that are extreme compared with most of the biosphere. For example, they may be extremely hot, cold, salty, acid, or polluted. In such habitats high populations of a few species are usually found. Presumably there are not more species because many kinds of organisms lack genetic capabilities for evolving the ability to tolerate a particular kind of extreme habitat. Here the resource range is narrow.

Resource Diversity. If the resources depended on by a particular trophic level or group of organisms are diverse, then that trophic level or group is likely to be diverse. The plant life of a physically diverse area—one with hills and valleys, rocky areas and good soil, wet spots and dry spots—will usually be more diverse than in a topographically uniform area such as a plain. There tend to be more bird species in vegetation with several layers such as a forest than in areas with little stratal diversity such as a grassland.

The area where two ecosystems come together, the transition zone between them, is called an *ecotone*. Species diversity of such areas tends to be high, and the main reason for this is clear. Where forest grades into grassland, the ecotone will have some characteristic forest species, some grassland species, and some additional species that require resources from both kinds of vegetation (food from the grassland and escape cover from the woody growth, for example). Members of the last group, characteristic of the deciduous forest-grassland ecotone, are called *forest-edge species*. Some examples are bobwhite, song sparrow, gray catbird, fox squirrel, cottontail, and white-tailed deer.

Many game animals are edge species; Aldo Leopold coined the term "edge effect" to refer to the increased density of such ecotonal species at edges of communities or in areas where two or more communities are intermixed in patches. "Edge effect" is now commonly used to refer to the increased diversity of ecotones, as well as to the increased densities that are often found. Increasing edge is a powerful tool in wildlife management.

Productivity. Generally more species can exist in areas of high productivity than low, the main reason being that in an unproductive area some features of an ecosystem—such as a certain kind of food—will be at too low a level to support a population of a species permanently. In a productive area the same feature is more abundant, becoming a resource that one or more species can use. The narrow niche becomes able to support a population of viable size.

Although this relationship between productivity and diversity seems typical of undisturbed ecosystems, the effect of increasing production artificially, such as by fertilizing a pond, is often just the reverse—species diversity is reduced. There may be several reasons for this but probably the most important is a disturbance of the balance between competitive species.

Climatic Stability. Areas of stable or predictable climate tend to have more species than those of unpredictable climate. Many species with very low populations can survive in an area of stable climate, whereas chance weather events, such as a cold day in June or an unusually icy winter, may kill all the members (since there are not very many) of such species in areas of unpredictable climate. Also, organisms in a stable climate can have narrower tolerance ranges, which may allow narrower niches.

Predation. Predators may keep numbers of certain of their prey low enough that competitive exclusion does not occur even though the prey species show considerable niche overlap. The best known example of predation keeping diversity high is the work of R. T. Paine. In the rocky intertidal zone of the Olympic Peninsula in Washington the top carnivore is a starfish that preys on large invertebrates such as mussels and barnacles. When the starfish was experimentally removed, several species disappeared as a mussel that is a preferred food of the starfish came to dominate the area.

The Shannon-Wiener Index

Species richness may be compared between two communities or areas by simply counting the number of species either in the whole community, if that is practical, or in suitable samples. However, there is at least one more aspect of species diversity that should sometimes be considered— relative abundance, or *equitability*. Imagine two communities each made up of two species. In one community there are 99 individuals of species A and one individual of species B, whereas in the second community there are 50 individuals of each species. The second community is more diverse in this sense: almost every individual sample is predictable in the first community (it will be species A) but not in the second.

One widely used measure of diversity that combines species richness with equitability is the **Shannon-Wiener index** (also called the Shannon-Weaver index). It is supposed to measure the uncertainty involved in predicting the identity of the next individual. The formula for calculating the index H' is

$$H' = -\sum_{i=1}^{S} P_i \log P_i$$

where S = number of species and P_i = the proportion belonging to the ith species. The logarithms used can be to any base. The base 2 is used by some because the answer H' is then in bits; however, most people use base e because it is easy and not so far from 2.

Calculation of H' may be illustrated using some fabricated numbers from tropical African savanna:

Species	No. of individuals	P_i	$\log_e P_i$	$P_i \times (\log_e P_i)$
Zebra	50	0.556	−0.5870	−0.326
Hyena	30	0.333	−1.0996	−0.366
Lion	10	0.111	−2.1982	−0.244
Total	90	1.00	—	−0.936

In the formula we take the negative sum of $P_i(\log_e P_i)$ for the practical reason of preferring to deal with positive numbers. H' for this set of data is then 0.936. (If the index is needed in binary form, the value $\log_e P_i$ may be converted to $\log_2 P_i$ by dividing it by 0.6931. For example, $\log_2 0.556$ equals $-0.5870/0.6931$ equals -0.8469.)

One practical application of the index has been in assessing stream pollution. For the bottom fauna an H' greater than 3 (using $\log_2$) usually indicates no, or very slight, pollution. Because some species decline or disappear while a few species increase, the index drops with pollution. An index of 2 to 3 indicates light pollution and an index below 1 indicates heavy pollution.

WHAT DETERMINES THE NUMBER OF SPECIES ON ISLANDS?

Biologists have known for a long time that an island will contain fewer species than an area the same size on the nearest mainland. It was thought that islands had an impoverished flora or fauna because of the difficulties of dispersal over long stretches of water. This idea may have some validity for a few kinds of organisms such as frogs and salamanders which are very poorly suited for long distance transport over seawater. But recent studies have shown that disseminules of most kinds of organisms arrive on islands at a high rate, although many of the species do not become established.

The island biologist Sherman Carlquist has included as one of his "Principles of dispersal and evolution" the statement that for species having the means of long distance transport the eventual introduction to an island is more probable than its nonintroduction. The primary reason for the lower number of species on islands seems to be that they do not have an overflow from adjacent areas in the same way as a mainland area.

There are other factors, however, including the fact that small islands may be too small to provide a large enough energy base to allow some of the top carnivores to exist.

R. H. MacArthur and E. O. Wilson have developed a model to explain how the number of species occupying an island is determined. They have suggested that the number is the result of a balance between immigration and extinction rates. They theorized that immigration rate should decrease and extinction rate should increase as the number of species on an island increases.

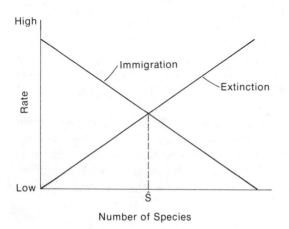

Imagine an island with certain traits having no species on it; immigration rate will be high because every species that lands as seed, spore, or adult will be a new one. Extinction rate will be low because there will be no species to become extinct. As species number increases, as it will if we started at zero, immigration rate will decline, since a larger and larger fraction of the possible species, or species pool, is already present. Extinction rate will increase because there are more species to become extinct and also because of increased interspecific competition (and possibly in some cases increased predation and disease). The number of species at the point where the immigration rate and the extinction rate are equal will be the equilibrium species number ($\hat{S}$, which is read "S hat").

Islands near a mainland as compared with islands further away should have a larger number of species because the immigration rate will be higher.

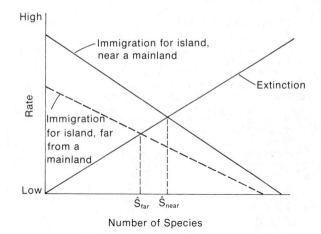

Number of Species

Also, large islands should have a larger number of species because extinction rate should be lower. This is presumably due to the greater diversity of habitats on a larger island; however, mere size may be important in providing a great enough energy base for some top carnivores to maintain populations large enough to avoid extinction.

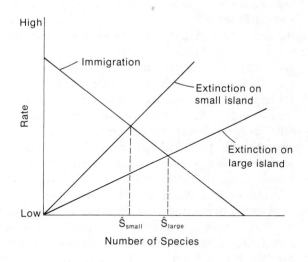

Number of Species

There are now several field studies which support the basic features of MacArthur and Wilson's island model. The Channel Islands off the California coast each had about the same number of bird species in 1968 as in 1917, but the identity of some 30% of the species was different. In other words, the species number was being maintained by a balancing of extinctions and colonizations.

Table 4–8 THE EFFECT OF FOREST STAND SIZE ON TYPES OF BIRDS
LIVING THERE*

Because of the way our land was surveyed, in townships, sections, and quarter-sections, isolated woodlots of 40 acres are frequent. Several of the species in the third column are rare in or absent from such stands. The species in the middle column are about equally likely to be in a forest of, for example, 20 acres as one of 60, but as the size of the woodlot decreases to 10 or 5 or 2 acres even these begin to drop out. About the only species that may be more common in smaller forests are the forest-edge species given in the first column.

Species More Likely to be Found in Forests of 35 Acres or Less	Species Not Much Affected by Forest Size	Species More Likely to be Found in Forests of 60 Acres or More
Downy woodpecker	Eastern wood pewee	Red-bellied woodpecker
House wren	Black-capped chickadee	Acadian flycatcher
Blue jay	White-breasted nuthatch	Wood thrush
Indigo bunting	Red-eyed vireo	Blue-gray gnatcatcher
Starling	Northern oriole	Cerulean warbler
	Brown-headed cowbird	Overbird
	Rose-breasted grosbeak	American redstart
		Scarlet tanager

*Adapted from material in Bond, R. R., "Ecological distribution of breeding birds in the upland forests of southern Wisconsin," *Ecol. Monogr.*, 27:351–384, 1957, and Galli, A. E., Leck, C. F., and Forman, R. T. T. "Avian distribution patterns in forest islands of different sizes in central New Jersey," *Auk*, 93:356–364, 1976.

D. H. Simberloff and E. O. Wilson censused the arthropods of several very small mangrove islands (10 to 20 meters in diameter) in the Florida Keys and then had the islands fumigated with methyl bromide, killing the arthropods but leaving the mangroves intact. They then followed the recolonization process. Within 200 days species number had returned to about the same as that found in "predefaunation" surveys.

An isolated beech-maple forest surrounded by cropland, or in pre-settlement times a prairie surrounded by forest, share some features with true islands. These situations are sometimes known as *habitat islands*. Small isolated forest plots often lack some of the characteristic forest species. As you go from large to small forest stands perhaps from 160 to 20 acres you lose pileated woodpeckers and Acadian flycatchers and gain house wrens and song sparrows (Table 4–8). There seem to be two likely reasons for this. First, the slow reproducing climax forest species may have extinction rates too high and immigration rates too low to be able to persist in isolated stands. The other likelihood is that competition from forest-edge species overflowing from the surrounding countryside may produce high extinction rates. Whatever the explanation, the implication is that preserving the climax forest fauna will require large preserves as well as small ones. In fact, the small ones may be of little importance for preserving many characteristic forest species.

BIBLIOGRAPHY

Baer, J. G. *Ecology of Animal Parasites*. Urbana, University of Illinois Press, 1951.
Berkner, L. V., and Marshall, L. C. "The history of growth of oxygen in the Earth's atmos-

phere," *in* D. J. Brancazio and A. G. W. Cameron, eds., *The Origin and Evolution of Atmospheres and Oceans.* New York, John Wiley & Sons, 1964, pp. 102–124.

Bond, R. R. "Ecological distribution of breeding birds in the upland forests of southern Wisconsin," *Ecol. Monogr.*, 27:351–384, 1957.

Brown, Frank A., Jr., Hastings, J. W. and Palmer, J. D. *The Biological Clock—Two Views.* New York, Academic Press, 1970.

Buechner, H. K. and Golley, F. B., 1967. "Preliminary estimation of energy flow in Uganda kob," *in* K. Petrusewicz, ed., *Secondary Productivity of Terrestrial Ecosystems,* 2 vols. Warsaw, Institute of Ecology, Polish Academic of Science, 1967, pp. 243–254.

Carlquist, Sherman. *Island Biology.* New York, Columbia University Press, 1974.

Cloud, P., and Gibor, A., "The oxygen cycle," *Sci. Am.*, 223(3):110–123, 1970.

Cody, M. L., and Diamond, J. M., eds. *Ecology and Evolution of Communities.* Cambridge, Harvard University Press, 1975.

Cummins, K. W., and Wuycheck, J. C. "Caloric equivalents for investigations in ecological energetics," *Int. Ver. Theor. Angew. Limnol. Verh.*, 18:1–158, 1971.

Darnell, R. "Organic detritus in relation to the estuarine ecosystem," *in* G. H. Lauff, ed., *Estuaries.* Washington, D.C., American Association for the Advancement of Science, 1967, pp. 376–382.

DeVries, D. A., and Afgan, N. H., eds. Heat and Mass Transfer in the Biosphere., Washington, Scripta, 1975.

Diamond, J. M. "Avifaunal equilibria and species turnover rates on the Channel Islands of California," *Proc. Natl. Acad. Sci.*, 64:57–63, 1969.

Eckardt, F. E., ed. *Functioning of Terrestrial Ecosystems at the Primary Production Level.* Paris, UNESCO, 1968.

Ehrlich, Paul R., and Raven, Peter H. "Butterflies and plants: a study in coevolution," *Evolution*, 18:586–608, 1965.

Engelmann, M. D. "The role of soil arthropods in the energetics of an old field community." *Ecol. Monogr.*, 31:221–238, 1961.

Forbes, S. A. "On some interactions of organisms," *Bull. Ill. Lab. Nat. Hist.*, 7:3–17, 1880.

Galli, A. E., Leck, C. F., and Forman, R. T. T. "Avian distribution patterns in forest islands of different sizes in central New Jersey," *Auk*, 93:356–364, 1976.

Gaufin, A. R. "The effects of pollution on a Midwestern stream," *Ohio J. Sci.*, 58:197–208, 1958.

Gilbert, L. E., and Raven, Peter H., eds. *Coevolution of Animals and Plants.* Austin, University of Texas Press, 1975.

Golley, F. B. "Energy flux in ecosystems," *in* J. A. Wiens, ed., *Ecosystem Structure and Function.* Corvallis, Oregon State University Press, 1972.

Heichel, G. H. "Agricultural production and energy resources," *Am. Sci.*, 64:64–72, 1976.

Henry, S. M., ed. *Symbiosis.* New York, Academic Press, 1966.

Hopkins, A. D. "Periodical events and natural law as guides to agricultural research and practice," U.S. Department of Agriculture, *Monthly Weather Rev. Suppl.* 9:1–42, 1918.

Hutchinson, G. E. "Concluding remarks," *Cold Spring Harbor Symp. Quant. Biol.*, 22:415–427, 1957.

———. "Homage to Santa Rosalia, or why are there so many kinds of animals?" *Amer. Nat.*, 93:145–159, 1959.

———. *The Ecological Theater and the Evolutionary Play.* New Haven, Yale University Press, 1965.

Jansen, D. H. "Coevolution of mutualism between ants and acacias in central America," *Evolution*, 20:249–275, 1966.

Kale, Herbert W. "Ecology and bioenergetics of the long-billed marsh wren *Telmatodytes palustris griseus* (Brewster) in Georgia salt marshes," Cambridge, Mass., Publ. Nuttall Ornith. Club, No. 5, 1965.

Kendeigh, S. Charles. "Bird population studies in the coniferous forest biome during a spruce budworm outbreak," *Ontario Dept. Lands and Forests Biol. Bull.*, 1:1–100, 1947.

Leak, W. B. "Some effects of forest preservation," U.S. Forest Serv. Research Note NE-186:1–4, 1974.

Leopold, Aldo, and Jones, S. E. "A phenological record for Sauk and Dane Counties, Wisconsin, 1935–1945," *Ecol. Monogr.*, 17:81–122, 1947.

Likens, G. E., and Bormann, F. H. "Nutrient cycling in ecosystems," *in* J. A. Wiens, ed., *Ecosystem Structure and Function.* Corvallis, Oregon State University Press, 1972.

Likens, Gene E., Bormann, F. Herbert, Pierce, Robert S., Eaton, John S., and Johnson, Noye M. *Biogeochemistry of a Forested Ecosystem.* New York, Springer-Verlag, 1977.

Lindeman, Raymond L. "The trophic-dynamic aspect of ecology," *Ecology*, 23:399–418, 1942.

Lovelock, J.E. and Margulis, L., "Atmospheric homeostasis by and for the biosphere: the Gaia hypothesis," *Tellus*, 26:2–10, 1974.

MacArthur, Robert H. "Patterns of species diversity," *Biol. Rev.*, 40:510–533, 1965.

———, and Wilson, E. O. *Theory of Island Biogeography*. Princeton, N.J., Princeton University Press, 1967.

Margalef, Ramon. "Information theory in ecology," *Gen. Syst.*, 3:36–71, 1958.

McDougall, W. B. "The classification of symbiotic phenomena," *Plant World*, 21:250–256, 1918.

Muller, C. H. "The role of chemical inhibition (allelopathy) in vegetational composition," *Bull. Torrey Bot. Club*, 93:332–351, 1966.

Odum, Howard T. 1957. "Trophic structure and productivity of Silver Springs, Florida," *Ecol. Monogr.*, 27:55–112, 1957.

———, and Pigeon, R. F., *A Tropical Rain Forest. A Study of Irradiation and Ecology at El Verde, Puerto Rico*. Springfield, Va., National Tech. Information Service, 1970.

Ovington, J. D. "Some aspects of energy flow in plantations of *Pinus sylvestris*," *Ann. Bot. n.s.*, 25:12–20, 1961.

Paine, R. T. "Food web complexity and species diversity," *Amer. Nat.*, 100:65–76, 1966.

Park, O. "Nocturnalism—the development of a problem," *Ecol. Monogr.*, 10:485–536, 1940.

Petrusewicz, K., ed. *Secondary Productivity of Terrestrial Ecosystems*, 2 vols. Warsaw, Institute of Ecology, Polish Academy of Science, 1967.

Pianka, E. R. "Latitudinal gradients in species diversity: a review of concepts," *Amer. Nat.*, 100:33–46, 1966.

Pielou, E. C. "The measurement of diversity in different types of biological collections," *J. Theor. Biol.*, 13:131–144, 1966.

Pimentel, David. "Species diversity and insect population outbreaks," *Ann. Entomol. Soc. Amer.*, 54:76–86, 1961.

———, et al. "Food production and the energy crisis," *Science*, 182:443–449, 1973.

Reichle, David E., ed. *Analysis of Temperate Forest Ecosystems*. Heidelberg, Berlin, Springer-Verlag, 1970.

———, Franklin, J. F., and Goodall, D. W. "Productivity of world ecosystems," *U.S. Natl. Acad. Sci. Symp.*, 1975.

Richards, P. W. *The Tropical Rain Forest. An Ecological Study*. New York, Cambridge University Press, 1952.

Riddiford, L. M., and Williams, C. M. "Volatile principle from oak leaves: role in sex life of the polyphemus moth," *Science*, 155:589–590, 1967.

Rodin, L. E., and Bazilevich, N. I., *Production and Mineral Cycling in Terrestrial Vegetation*. Edinburgh, Oliver and Boyd, 1967.

Simberloff, D. S., and Wilson, E. O. "Experimental zoogeography of islands: the colonization of empty islands," *Ecology*, 50:278–296, 1969.

Sondheimer, E., and Simeone, J. B., eds. *Chemical Ecology*. New York, Academic Press, 1969.

Stanchinsky, V. V. "On the importance of biomass in the dynamic equilibrium of the biocenose," (in Russian), *J. Ecol. Biocenol*, 1:88–98, 1931.

Steele, J. H., ed. *Marine Food Chains*. Berkeley, University of California Press, 1970.

Steinhart, J. S., and Steinhart, C. E. "Energy use in the U.S. food system," *Science*, 184:307–316, 1974.

Teal, John M. "Community metabolism in a temperate cold spring," *Ecol. Monogr.*, 27:283–302, 1957.

Transeau, E. N. "The accumulation of energy by plants," *Ohio J. Sci.*, 26:1–10, 1926.

Van Hook, R. I. "Energy and nutrient dynamics of spider and orthopteran populations in a grassland ecosystem," *Ecol. Monogr.*, 41:1–26, 1971.

Vitousek, P. M., and Reiners, W. A. "Ecosystem succession and nutrient retention: a hypothesis," *BioScience*, 25:376–381, 1975.

Wald, G. "The origins of life," *Proc. Natl. Acad. Sci.*, 52:595–611, 1964.

Westlake, D. F. "Comparison of plant productivity," *Biol. Rev.*, 38:385–425, 1963.

Whittaker, R. H., and Feeny, P. P. "Allelochemics: chemical interactions between species," *Science*, 171:757–770, 1971.

Wilde, S. A. "Mycorrhizae and tree nutrition," *BioScience*, 18:482–484, 1968.

Chapter 5

COMMUNITY AND ECOSYSTEM ECOLOGY: COMMUNITY CHANGE AND THE NATURAL LANDSCAPE

COMMUNITY CHANGE

Communities and ecosystems are dynamic; they change constantly. One tree dies and a sapling will grow up to take its place. A forest is a different place from night to day and from spring to fall. Winter comes as a prelude to another season of growth. Such changes as these, however, do not alter the community permanently; they are called *nondirectional* changes. These include the replacement changes associated with maintaining a steady state in stable communities. Several kinds of cyclic and periodic changes are also included; the most obvious are the daily cycle of light and dark and the yearly cycles of temperature, photoperiod, and rainfall. *Fluctuations* in communities associated with climatic fluctuations, such as changes in grassland resulting from a series of drought years, are non-directional changes.

Some other types of community change do result in permanent alteration of the community. They are *directional* changes. Included are the long-term changes resulting from long-term climatic change, such as the shifts of plants and animals that occurred with glacial advance and recession. Changes that occur on even a longer time scale which involve the evolution and extinction of species, such as the changes in land communities as birds and mammals evolved, are directional. So also is *succession*, the process of community development which may produce a climax community (see p. 178). It is a directional change which occurs even though climate and the species pool available do not change

The community, of course, has no regard for ecologists' categories.

175

The decrease of a certain xerophyte may be part of a climatic fluctuation in which rainfall is higher, part of a long-term climatic trend toward greater rainfall, or part of succession toward a more mesic community. All three of these types of community change may be happening at once.

Fluctuation, succession, and long-term community change (under *Paleoecology*) are discussed in the following sections of this chapter. Daily and seasonal change have been dealt with in Chapter 4 in the section on community organization.

THE GREAT DROUGHT: AN EXAMPLE OF FLUCTUATION

One of the best studied examples of the "fluctuation" type of community change was the change in grassland vegetation that occurred during and following the Great Drought of the 1930's in the Great Plains. The effects of the drought were studied by the grassland ecologist John Weaver and his coworkers in Nebraska, Iowa, and Kansas. Rainfall for six years preceding 1933 had been 5 inches above average; during the early years of the drought there was a deficit of 7 inches below the long-term average of 23 inches a year. The percentage of ground covered by vegetation in ungrazed short-grass prairie was about 85% before the drought. This percentage steadily declined as the grasses died back or failed to come up in the spring—it was 65% by 1935, 30% in 1938, and 20% in 1940. Litter disappeared, decomposing faster than it was added.

Not all species of plants declined. The cover of little and big bluestem dropped from 61 to 3% between 1932 and 1939, but the grama grasses increased from a negligible percentage to 27%. If these arid-land grasses from the Southwest had not increased the ground would have been even more bare. Another grass that increased was western wheatgrass, a plant that makes most of its growth in the spring when moisture is most available and which has a high percentage of its roots deep in the soil. Some of the shallowly rooted forbs* such as coneflower and pussy-toes disappeared in the first two years of the drought, but several forbs, such as some of the asters and goldenrods, increased for a while as competition from the grasses declined.

With the ground bared soil began to blow (Fig. 5–1). Blowing of topsoil was worst in the southwestern part of the Great Plains where Oklahoma, Kansas, Texas, Colorado, and New Mexico come together. Row-crop farming, mainly wheat, had expanded in this region onto lands that should have been left in their native short grasses. With the coming of the drought this became the Dust Bowl, an area of 50 million acres by 1936 from which farm families ruined by crop failures began to flee,

---

*A *forb* is a herb that is not a grass, sedge, or rush.

Figure 5–1. An approaching dust storm in Texas, August 1, 1936. (Photo courtesy U.S. Soil Conservation Service.)

mostly to the "Promised Land" of California. John Weaver studied one face of the drought; the story of another was told by John Steinbeck in *The Grapes of Wrath* and by Woody Guthrie in such songs as "Dust Bowl Refugees" and "Dust Pneumonia Blues." Their writings are less scientific than Weaver's but of no less ecological interest.

It was not until the winter of 1941–42 that the drought really ended. Rainfall increased, and it was found that many grasses still lived as roots and underground stems, even though they had not shown themselves aboveground for several years. Now they sprouted, cover and shade increased, and by 1943 a mulch of leaves and stems from the preceding year began to cover the ground again. Forbs also began to sprout from underground stems or from seeds that had lain dormant in the soil. The forbs increased steadily and in 1945 and 1946 several species were growing in sites that the grasses had held before the drought. Not until 1959–1960 did forb distribution on Weaver's study areas return to its predrought condition. Western wheatgrass, one of the xerophytic grasses, was the slowest to decline because its early spring growth allowed it to resist encroachment by other grasses. Shading by the taller bluestems finally caused its decline.

One trap that both ecologists and farmers must avoid is thinking of the wet years just preceding the 1930's as the normal condition. Drought, no less than years of good rain, is normal for these semi-arid regions. Droughts had come before—an especially bad one in the 1890's gave rise to the slogan "In God we trusted, in Kansas we busted"—and they came afterwards (the Dust Bowl area experienced another drought in the mid-1950's). They will come again. Land use in such an area must be designed for living with dry as well as moist periods. An ecologist studying the area must realize that western wheatgrass is as much a member of the ecosystem studied by Weaver as is big bluestem.

SUCCESSION

If a new area essentially free from life is produced, for example by a river depositing a new sandbar or by humans strip-mining an area for coal, the community that initially develops there is a *pioneer community*, which usually does not persist very long. In the course of time various species are lost and others invade. These, in turn, may disappear and still others may enter. Some tens, hundreds, or thousands of years later, a community will have developed which is stable or at a steady-state, known as the *climax community*. The process of development from pioneer to climax community is referred to as *succession*.

Pioneer communities tend to be made up of relatively few species, those that are able, first, to immigrate quickly and, second, to live under the particular environmental conditions, often extreme, of the new area. The earliest plants to grow may be those with either wind-blown seeds or those that are carried long distances in or on birds or mammals; they would be plants that could withstand direct sunlight and various other harsh conditions. Legumes, able to grow in nitrate-poor soil because of the nitrogen-fixing bacteria associated with their roots, may be prominent. For the first one or two years annual herbs may be most important, but they are soon replaced by perennial herbs. One obvious factor involved in this replacement is that of occupancy. Once a perennial has established itself it is able to hold onto an area and, usually, to spread by vegetative means. Because annuals have to grow from seed each year they are usually crowded out quickly.

This perennial herb state might be replaced by a community dominated by shrubs, followed by a community dominated by a forest of light-tolerant trees. Some of these changes represent differences in the time of invasion of various species, while others represent merely the different time needed to reach dominance. The trees may invade about the same time as the shrubs but for several years they are no taller; later they begin to overtop the shrubs and change the look of the community.

The light-tolerant trees are also generally shade-intolerant and are unable to reproduce in the shade they cast. Beneath them shade-tolerant trees invade, and of course there are also shifts from shade-intolerant to shade-tolerant shrubs and herbs. At this point we may have a community in which the canopy trees are mainly shade-intolerant species whereas all the lower layers are shade-tolerant species. The seedlings of shade-tolerant trees grow up and, as the shade-intolerant trees die, a community of shade-tolerant trees develops. The community is at or near climax.

The whole sequence from bare ground to climax community is called a successional series or a *sere*, with the various communities existing at different points in time called *seral stages*. For the sere just discussed the following seral stages might be recognized: annual herb, perennial herb, shrub, early forest, followed by the climax forest.

The causes of succession may be classified into two categories, *allogenic* and *autogenic*. Allogenic causes are those exterior to the community. In succession on bare rock (a very slow process, by the way) some of the breakdown of the rock is by organisms but some is by physical and chemical weathering. These latter processes would be allogenic. In pond succession the filling-in by sand, silt, and clay carried from surrounding lands is an allogenic factor.

However, autogenic changes are particularly important in succession, and *reactions* (Chap. 4) are especially significant. F. E. Clements has spoken of reactions as being the driving force of succession. Generally in the course of time the reactions of the dominant organisms of each stage make the area less favorable for themselves and more favorable for organisms of the next stage. We should be clear that this is a matter of invasion and dominance by organisms able to take advantage of changes caused by the earlier species, and not a matter of the earlier species having as their "function" paving the way for the later ones; there is no evidence that ragweed is altruistic.

The process continues to the climax, which is composed of organisms that can tolerate their own reactions. This is seen in the sequence of more and more shade-tolerant organisms paralleling the deeper and deeper shade produced during the course of the sere just described, and in the addition of nitrate to the soil by plants with associated nitrogen-fixers allowing the invasion of organisms with higher nitrate requirements.

Other important autogenic factors are immigration, growth rates, and coactions. Competition leading to the elimination of less nearly climax species by more nearly climax species is probably an important coaction. Other, noncompetitive, coactions are also important—such as trees providing nesting sites for forest birds, or the arrival of earthworms and snails providing food for shrews.

Succession may begin in many different ways. In the examples discussed so far it started on a totally bare area, an area that had never before supported a community. When succession occurs in such a place, on a new river bar or a lava flow, it is called *primary succession*. *Secondary succession* occurs on areas which have already supported a community, for instance on abandoned cropland allowed to revegetate or on a piece of burned–over forest land. Usually the rate of secondary succession is faster than that of primary succession because there are some organisms and a developed soil already present.

Within any community, seral or climax, particular microhabitats may support a sequence of populations or communities making up a *microsere*. It is difficult to give an absolutely satisfactory definition of microsere but the idea is easily grasped. A forest tree blows over. The sequence of changes which include the various insects and other organisms involved as the tree disintegrates, decays, and eventually becomes indistinguishable from the general forest floor is one microsere. Other microseres that have been studied occur on the dead bodies of animals (Table 5-1), on cattle (or bison) droppings, in seasonal ponds, and in pond infusion cultures.

Sand Dune Succession

Along its southern and eastern shores Lake Michigan is surrounded by sand deposited by the lake waters and thrown up into dunes by the action of the wind. These sand areas extend inland a considerable distance from the present-day shortline because at times during the period when the glaciers were retreating the lake level was higher than now. To the casual visitor the distribution of communities on the sand areas may seem chaotic, but there is a pattern, worked out by Henry Chandler Cowles and Victor E. Shelford in some of the earliest ecological studies done in the U.S.

The area adjoining the lake is divided into the lower, middle, and upper beaches (Fig. 5–2). The lower beach includes the area washed by summer storms. The middle beach is the section from this point to the limit of wave action of winter storms. A line of driftwood and debris usually marks the upper edge of the middle beach.

Because of its instability the lower beach supports no vegetation. However the water-filled spaces between the sand grains a few centimeters below the surface may support bacteria and a large invertebrate fauna of protozoa, rotifers, copepods, and tardigrades. The middle beach provides a relatively stable area during the summer, with the result that annual plants are able to grow and reproduce here. Perennial plants may start growth in one summer but tend to be washed away or buried under new sand in winter. The annuals which exploit this specialized habitat are succulents, a trait usually associated with deserts. In fact, the middle beach is desert-like despite the abundance of soil moisture. The high temperatures near the surface and the constant winds make loss of water by evaporation high enough that water balance is a problem for organisms living here. Some of the succulent annuals occurring on the middle beach are sea rocket, bugseed and a spurge.

Table 5–1 SOME FEATURES OF THE CARRION MICROSERE*
In an acre of forest (or grassland or marsh) several fair-sized animals die every day. Relatively few of these are seen because they are quickly eaten by opossums or vultures or buried by those remarkable insects called burying beetles. If this does not occur the carcass goes through a series of changes like those described below. The particular study on which this table is based was done by M. D. Johnson, who put out medium-sized carcasses (mostly squirrels) in hardware-cloth cages in a red oak-sugar maple-basswood forest in northern Illinois.

Stage	Duration (Spring and Summer)	Characteristics
Fresh	3 hours–2 days	Blowflies may visit the carcass to lay eggs within hours or even minutes of death.
Bloat	4–19 days	Begins when the carcass bloats from the gases produced by anaerobic decomposition within. Blowfly maggots hatching from eggs feed on the flesh. Predaceous beetles such as rove and hister beetles enter the carcass to prey on the blowfly larvae.
Decay	13–23 days	The burrowing of maggots and other insects allows the entrance of air into the carcass, halting putrefaction and allowing aerobic decomposition to occur. The maggots contribute to decomposition by secreting digestive enzymes; bacteria and fungi are also important in decay. This is the period of greatest insect abundance—the carcass may swarm with fly larvae. Blowflies peak about the beginning of this stage and then leave the carcass to pupate; however, other kinds of scavenging flies remain. Beetles, both scavengers and predators, are common during this stage.
Dry	30–60 days	By the end of the decay stage the carcass is a hollow shell of hard, dry skin. The lower side is disintegrated and hair and bones are scattered around the carcass. Insect populations are much diminished; some fly larvae of a few kinds remain, but beetles, mainly scavengers, are more numerous. An example is the skin beetle, so called because of its occurrence at this stage. By the end of this stage characteristic carrion species are gone and the invertebrates found are typical soil and litter forms such as millipedes and daddy longlegs.

*From Johnson, M. D., "Seasonal and microseral variations in the insect populations on carrion." Am. Midl. Nat., 93:79–90, 1975.

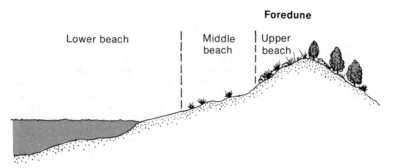

Figure 5–2. Zonation on a Lake Michigan beach. The lower beach, washed by the waves of summer storms, supports no plants. Sparse vegetation, mostly annuals, grows on the middle beach. The upper beach, outside the reach of the highest waves of winter, supports permanent vegetation, such as marram grass and cottonwood.

Relatively few animals live permanently on the middle beach. Most of these are either scavengers or predators on the scavengers. The scavengers make their living off the bodies of animals washed up on shore. Some of these are fish, as might be expected, but there are also large numbers of insects carried over the water, drowned and washed onto shore, and also, during spring and fall, migrating birds which are killed by storms or become exhausted while over the lake. The scavengers include such invertebrates as blowflies, dermestid and rove beetles, and also vertebrates such as opossums and grackles which come to the beach from other inland communities. Tiger beetles and spiders are predators on the invertebrate scavengers.

The lower and middle beaches are ecologically interesting but their biotas are totally dependent on the lake; consequently they are not really part of the sand dune sere. Only if the lake level were to drop for several years would the current lower or middle beaches begin to undergo successional changes such as described here beginning with the vegetation of the upper beach.

The upper beach is out of reach of the waves. Typically we find a dune, the *fore-dune*, occupying the upper beach; the dune is built as a result of plant activity. Occasionally, through some catastrophic cause, a flat area will be produced in the upper beach zone and, here *embryonic* dunes will begin to form.

Dunes are built by sand being deposited around something which presents an obstacle to the wind; the wind is slowed and consequently drops the sand it is carrying. Small, temporary dunes can be formed by snow fences or even by a pile of trash but only plants are able to build large, permanent dunes. There are three traits that make a plant a good dune former: (1) it must be perennial so that it holds the sand from year to year; (2) it must be able to grow upward, surviving continual burial as sand is deposited around it; and (3) it must be able to spread laterally so that the dune can grow in width as it grows in height.

Two species of grass, beach or marram grass and dune grass, are the best dune formers and generally occupy much of the fore-dune (Fig. 5–3). Both are perennial, can put out new roots from buried stems while sending up new shoots above the sand, and can send out long underground stems, *rhizomes*, from which new aboveground stems sprout. Some other grasses and a few woody plants are also important members of the flora of the upper beach. Possibly the most important dune-forming woody plant is a small shrub, the sand cherry, that forms large patches by sprouting from its spreading roots. Cottonwood is also important; it can survive burial by continuing to grow upward and putting out new roots but it does not spread laterally as do the grasses and sand cherry.

If we look at succession beginning at an embryonic dune being formed around a clump of grass, we can consider the pioneer stage in the sere to be a herbaceous stage dominated by the plants mentioned above and some others which generally invade, although they contribute little to dune formation. One is yellow beach thistle, a species that occurs only on the sand dunes around Lakes Michigan, Huron, and Superior.

A fore-dune is built, with the sand stabilized by these plants. At this

Figure 5–3, 4. Sand dune succession. A marram grass-covered foredune is shown in A. With high lake levels, dunes that were deposited in years of lower levels may be eroded by wave action. In B, the waves have cut away a foredune, exposing the spreading rhizomes that make marram grass such an effective dune former. In C, a blowout has produced a wandering dune which is moving away from the lake (*left*) across an area of bearberry, juniper, and jack pine. In the sheltered valleys produced by the slopes of old dunes (D), basswood is one of the important tree species.

point succession may proceed through other stages (described below), and eventually a forest may occupy this fore-dune area. Sometimes, however, the dune covered by dune and beach grass may become a *wandering dune*; the vegetation loses its grip on the sand, and the wind begins to move the dune back across the landscape. What it is that causes a wandering dune to form is not completely clear, but some disturbance to the vegetation that allows the wind to get at the sand is usually involved. Such disturbances in the form of dune schooners, from which tourists can view the dunes without having to walk up and down hill, motorcycles, construction of houses, and many other human activities is probably much more prevalent than in primeval times. Usually the start of a wandering dune seems to be in the form of a *blowout*. Vegetation is destroyed, the wind blows the exposed sand out, undermining the trees at the edges of the blowout, which enlarges rapidly as the vegetation on all sides of the hole topples in through the action of gravity.

The wandering dune moves back across the landscape through what might be called the *dune complex*. It may cover pine, oak, or even beech-maple forests, and fill in ponds. Eventually a point is reached where the sand is no longer being moved very much by the wind. Reduction of the wind may occur partly through distance from the lake but the formation lakeward of other dunes, including a new fore-dune, is also important.

At this point the dune, which during its wanderings has been simply bare sand, can now be captured by vegetation again. On the windward side of the dune the invading vegetation tends to be similar to that of the upper beach. Marram and dune grass invade, along with several other plants. This stage may be replaced by a shrub stage in which patches of bearberry, junipers, and various cherries, dogwoods, and willows are intermixed with vegetation of the preceding stage.

A cottonwood stage may occur at this point but, when it does, cottonwood does not represent a newly invading species. Cottonwood germinates only on wet sand so that the usual site for a cottonwood to begin is on a blowout which removes sand all the way down to the water table. When cottonwood trees appear on a dune in succession, then, they have invaded long before and become dominant by being able to grow higher and shade the other plants.

If no cottonwoods have become established at an earlier time, the shrub stage is frequently invaded by jack pines, and occasionally other species of pines. Like the cottonwood stage the jack pine stage is an open savanna vegetation. There is enough light reaching the ground in either stage for oaks to germinate and grow. Around the south end of Lake Michigan a rather low, open black oak forest seems to be climax. The effects of black oak forest on the sand do not seem to favor the eventual invasion of beech, maple, or basswood. In certain protected pockets in this region, however, and generally further north along the eastern shore of the lake, oak forest (usually red oak) is invaded by basswood, sugar maple, beech, and hemlock. The climax forest here seems to contain all five of these species.

These seem to be some of the sequences occurring on the windward slope of stabilized dunes, sequences involving some hundreds of years. Succession is faster on leeward slopes. Physical conditions are more favorable, and usually the leeward slope is immediately adjacent to an already developed forest.

If we return now to a fore-dune occupied by marram grass and sand cherry, and suppose that it does not suffer any disturbance that might convert it to a wandering dune, it will undergo a succession similar to that just described. It is not unusual to find a section of forest-covered fore-dune next to one occupied by marram grass. But not even forested dunes are immune to blowouts; human disturbance or possibly a tornado may open up a section of forest enough so that even a dune occupied by beech-maple-hemlock forest may be converted into a wandering dune.

Within this basic framework, most of the communities of the dunes complex can be understood. The dunes complex is a mosaic of embryonic dunes, blowouts, wandering dunes, and stabilized dunes in all stages of succession from marram grass to beech-maple forest. Even this is not the whole story. There are some additional seres involved in the sand areas, such as those around the interdunal ponds.

The sand dune sere described is based on reactions of the dominant plants, mainly shading and soil development (Table 5–2). Other reactions that they produce include moderated temperature extremes, a decrease in wind, an increase in relative humidity, and decreased evaporative power

Table 5–2 PHYSICAL FEATURES OF SAND DUNE SUCCESSIONAL STATES*
Disregarding blowouts, successional changes in the dominant plants on the sand dunes are caused mainly by shading and soil development. These effects are reactions, exerted by the plants themselves. The plants also produce other effects on the physical environment.

	Successional Stage				
Physical Feature	Marram Grass	Cotton-wood	Jack Pine	Oak	Beech-Maple
Midsummer light intensity 4 inches from the ground (expressed as % of full sunlight)	96	—	37	2	1
Soil moisture (weight of water expressed as % of dry weight of soil)	1	—	2	5	24
Summer evaporation rate (ml water per day)	—	21	11	10	8
Soil pH	7–8	—	5.5–7.0	5.5–7.0	5.5–6.0

*Evaporation data from Fuller, G. D., "Evaporation and plant succession," *Bot. Gaz.*, 52:193–208, 1911; other data original.

of the air. Based on these and other changes, the subdominant plants and the animals also change throughout the sere.

Changes in mammal and bird populations are shown in Figure 5–5. Invertebrate populations change also. Soil ants are numerous in the early successional stages; in the beech-maple forest there are fewer species and these tend to be log and tree-inhabiting forms. Among the spiders burrowing forms are characteristic of the marram grass and cottonwood stages but web-building species increase as the amount of vegetation increases. The largest number of spider species is in the black oak forest, with the second largest number in the beech-maple forest. Among the orthoptera, the trend is from short-horned grasshoppers in the marram grass to crickets in the beech-maple forest. Although changes occur steadily in the sere, it is characteristic for the biggest turnover of species to occur between the jack pine and oak stages. This is true for all of the groups mentioned—ants, spiders, orthoptera, birds, and mammals.

THE CLIMAX COMMUNITY

The *climax community* is in equilibrium with its environment, and climate is an especially important factor. The climax is at a steady state of species competition, structure, and energy flow. "Steady state" indicates the dynamic nature of the climax; changes occur continually but they are changes that tend to perpetuate the community rather than alter it.

Figure 5–5. Succession of mammals and birds on sand dunes around Lake Michigan. Numbers are individuals per 10 hectares. For birds, asterisks in the marram grass and cottonwood stages indicate relative abundance; the figures given are for a mixed marram-cottonwood community. (Data from: J. Van Orman, "Avian succession on Lake Michigan sand areas," M.S. Thesis, Western Michigan University, 1976; and J. Olson, personal communication.)

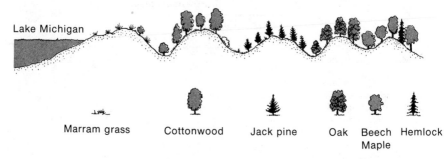

Species	Marram Grass	Cottonwood	Jack Pine	Black Oak	Beech-Maple-Hemlock
Prairie deer mouse	5	11			
13-lined ground squirrel	1	40	2		
Meadow jumping mouse		1			

Species	Marram Grass	Cottonwood	Jack Pine	Black Oak	Beech-Maple-Hemlock
Red squirrel			25		12
Eastern chipmunk			35	42	47
Masked shrew				2	4
Short-tailed shrew				1	15
Fox squirrel				15	10
White-footed mouse					3
Gray squirrel					2
Killdeer	** 3				
Field sparrow	* 10	**			
Red-winged blackbird	* 10	*	7		
Song sparrow	6	**	3		
Vesper sparrow	** 5	*	4		
Mallard	* 4	*	4		
Brown thrasher	3	**	7		
Eastern kingbird	* 2	**	3		
Prairie warbler		**			
House wren			3		
Chipping sparrow			2		
Mourning dove			4		
Northern oriole			4	4	
Cardinal			7		7
American robin			4	1	8
Blue jay			4	4	6
Great crested flycatcher			2	3	4
Black-capped chickadee			2	6	4
Blue-gray gnatcatcher				1	
Yellow and black-billed cuckoos				5	
Scarlet tanager				9	8
Ovenbird				4	10
White-breasted nuthatch				5	5
Tufted titmouse				3	4
Eastern wood pewee				6	10
Red-eyed vireo				7	11
Wood thrush				2	10
Downy woodpecker				2	4
Rose-breasted grosbeak				6	7
Common flicker				1	3
Hairy woodpecker				1	2
Acadian flycatcher					6
American redstart					11
Black-throated green warbler					4
Hooded warbler					1

Climax and other relatively mature ecosystems differ, at least in degree, from pioneer and other immature ecosystems. The following seem to be valid statements about the climax community, although individual exceptions may occur:

1. The climax community is able to tolerate its own reactions.

2. The climax tends to be mesic for the climate in which it occurs; many pioneer communities are relatively xeric or hydric.

3. The climax community tends to be more highly organized; for example, stratification may be more complex and there may be more, and more complex, coactions (more complex food webs, more mutualism).

4. Mature communities have more species (higher diversity) than immature ones, which tend to have high populations of a few species. This latter situation is to be expected where a relatively harsh environment permits only a few species to occur. The larger number of species in the climax community is related to the more complex organization, thus allowing more niches. However, it is now clear that there may be some loss of diversity in very old climax forests, as the species list drops to those best able to tolerate the low light and other conditions. Normally this loss of diversity is reversed at intervals by processes that "disturb" small patches of the forest—such things as severe wind storms and infrequent surface fires (Fig. 5–6). These processes rarely eliminate the climax species but allow some other species, perhaps ones less shade-tolerant, to persist or to invade again.

5. The organisms composing the climax community tend to be long-lived, relatively large, and with a low biotic potential (K-selected), whereas species of earlier successional stages tend to be smaller, shorter-lived, with a higher biotic potential (r-selected). Most croplands are man-made pioneer ecosystems, and most crops are annuals or short-lived perennials. Most game animals, such as bobwhite and rabbits, are species of relatively early successional stages; they produce a large number of young each year many of which, when not harvested by man, fall to predators and otherwise die. The mourning dove, for example, is a species

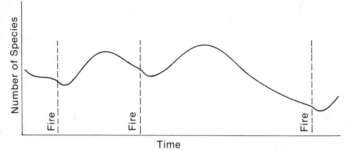

Figure 5–6. In this graphical representation of O. Loucks' model of species diversity relations in northern forests, "random perturbations" such as forest fires allow the growth and establishment of species which lose out in the stretches between perturbations. (Modified from O. L. Loucks, "Evolutions of diversity, efficiency, and community stability," *American Zoologist*, 10:23, 1970.)

of earlier successional stages. A given female lays only two eggs at a time but re-nests several times a year. Ecologically it is an appropriate game species, at least in the South. The related passenger pigeon, however, was a forest bird living on beechnuts and acorns and produced only a single egg a year. It should not have been subjected to hunting pressure and when it was (mainly market hunting) it quickly became extinct.

6. Gross primary production tends to be greater than community respiration in immature stages, whereas energy is at a steady state (net community production, NCP, is zero) in climax communities. This implies that energy is being stored in immature systems and, consequently, the amount of plant or animal tissue or dead organic matter (such as litter or humus) is increasing. In immature systems this excess can be harvested by man. A cornfield starts with seeds weighing pounds and ends up the season with ears and stalks weighing tons.

7. The stability of climax ecosystems is high. This is obviously true in the sense that immature ecosystems are temporary, but it also seems often to be true in the senses that immature ecosystems show broader changes than mature ones and are less resistant to outside disturbance.

In his thoughtful and thought-provoking essay entitled "The Strategy of Ecosystem Development," Eugene Odum discusses the relationship between ecosystem development and human use of the landscape. The following quotation makes some important points better than I can.

Man has generally been preoccupied with obtaining as much "production" from the landscape as possible, by developing and maintaining early successional types of ecosystems, usually monocultures. But, of course, man does not live by food and fiber alone; he also needs a balanced CO_2-O_2 atmosphere, the climatic buffer provided by oceans and masses of vegetation, and clean (that is, unproductive) water for culture and industrial uses. Many essential life-cycle resources, not to mention recreational and esthetic needs, are best provided man by the less "productive" landscapes. In other words, the landscape is not just a supply depot but is also the *oikos*—the home—in which we must live. Until recently mankind has more or less taken for granted the gas-exchange, water-purification, nutrient-cycling, and other protective functions of self-maintaining ecosystems, chiefly because neither his numbers nor his environmental manipulation have been great enough to affect regional and global balances. Now, of course, it is painfully evident that such balances are being affected, often detrimentally. . . . The most pleasant and certainly the safest landscape to live in is one containing a variety of crops, forests, lakes, streams, roadsides, marshes, seashores, and "waste places". . . As individuals we more or less instinctively surround our houses with protective nonedible cover (trees, shrubs, grass) at the same time that we strive to coax extra bushels from our cornfield. We all consider the cornfield a "good thing," of course, but most of us would not want to live there, and it would certainly be suicidal to cover the whole land area of the biosphere with cornfields. . . . The basic problem facing organized society today boils down to determining in some objective manner when we are getting "too much of a good thing." This is a completely new challenge to mankind because, up until now, he has had to be concerned largely with too little rather than too much. Thus, concrete is a "good thing," but not if half the world is covered with it.*

Science, 164:266–267, 1969. Copyright 1969 by the American Association for the Advancement of Science.

Mono- and Polyclimax

F. E. Clements was a clear observer who liked to fit complicated matters into systems. His view of succession and the climax community, now called the *monoclimax* hypothesis, was that within a given region all land surfaces eventually tend to be occupied by a single kind of community which is climax. The climax is determined by the regional climate. Given a stable climate the climax community is stable indefinitely.

There are several types of observations that seem to conflict with this hypothesis. For example, there is evidence that even under primeval conditions it was difficult to find large areas of uniform vegetation. In southwestern Michigan at the time of settlement there was about as much oak-hickory forest as beech-maple forest and there was also prairie, oak savanna, marsh, swamp forest, and sand barrens (Fig. 5–7). This area had about 15 or more thousand years (after the last glaciers) for the climax community to establish itself. Should we not recognize several different communities as climax on various sites (the *polyclimax* hypothesis)?

Clements's answer would probably have been that replaceability, not persistence, is the key to the status of a community. The drier, less shade-tolerant vegetation on the tops of hills may last for a very long time but as a mature soil develops it will be replaced by the climax vegetation; if not then, at least when the hills erode down further (as, of course, they surely must). It is clear to anyone, Clements would have said, that vegetation is a veritable mosaic, consisting of sites supporting the climax, sites that are undergoing rapid succession, and sites where succession is proceeding slowly or is stalled. The solution is not to tamper with the concept but to find appropriate names for the various nonclimax stages that are seen. Clements gave them names such as subclimax, serclimax, and disclimax (listed in Table 5–3).

Few present-day ecologists would accept this solution. They would argue that a climax community is one in harmony with the whole environment, not just climate. Climate is not very stable; the climatic climax in southern Michigan has, in the 15,000 years since the glaciers withdrew, been tundra, boreal forest, moist deciduous forest, and dry deciduous forest. Why, then, choose climate as the one factor determining *the* climax? There are also some other, more abstract, objections.

The solution of the polyclimax school is also basically terminological. After sorting out the obviously seral communities, they would recognize different kinds of climaxes, such as edaphic (where soil factors are important), topographic (elevation, slope), and fire.

A third hypothesis, called by R. H. Whittaker the *climax pattern concept,* rejects the classification approach. In this view the undisturbed vegetation of southwestern Michigan or the Smoky Mountains of Tennessee and North Carolina would be treated not by recognizing a bunch of communities but by studying and describing the spatial changes in the vegetation of the region. In the climax pattern concept there is, in a sense,

Figure 5–7. This map of southwestern Michigan shows that the primeval vegetation was far from uniform. The map was constructed from information contained in the notes of the original land survey. As land was opened up for settlement, government surveyors laid out the lines for township, range, and section. Part of the procedure involved marking and recording two or four trees at each section corner and halfway down some section lines. Consequently, there is information available on the forest or other vegetation at least every mile and sometimes at shorter intervals. (From L. A. Kenoyer, "Forest distribution in southwestern Michigan as interpreted from the original land survey (1826-32)," *Papers of the Michigan Academy of Science,* 19:108, 1934.)

Table 5-3 SOME CLEMENTSIAN TERMINOLOGY FOR COMMUNITIES
Some of these terms have become part of the everyday language of ecology. Others, such as serclimax and lociation, are now scarcely used.

Term	Interpretation
Climax	There is only one type of climax, that controlled by climate; accordingly *climatic climax* is redundant.
Proclimax	Any community that is persistent but lacks the proper sanction of climate. The following five terms are types of proclimaxes:
Subclimax	The successional stage preceding the climax, sometimes long persistent. Oak-hickory forest in a beech-maple climate is an example.
Serclimax	Persistent examples of stages earlier than the subclimax. The marshes of the Everglades are an example.
Disclimax	A community maintained by continued disturbance—for example, by overgrazing. Clements considered forests resulting from the elimination of the American chestnut by the chestnut blight as disclimax, possibly (I am not sure) because the blight was introduced by humans.
Preclimax and postclimax	Climaxes are spatially arranged, as can be seen in their occurrence as continental belts and as zones on mountains. Each climax is preclimax to the contiguous community of "higher" life form and postclimax to the contiguous community of "lower" life form. In the oak-hickory forest region, prairie preclimax might exist on well drained, west-facing slopes. Beech-maple forest (postclimax) might exist in moist, north-facing ravines.
Association	Associations are the primary divisions of the biome. Although the biome is recognized on the basis of the growth forms of the dominants, the different associations are separated by the kinds of plants composing them. The oak-hickory association would be one of the subdivisions of the temperate deciduous forest biome.
Faciation	This is a subdivision of an association corresponding to a particular subregional climate. In the mixed prairie association, three faciations are recognized; there is, for example, a southwestern faciation, characterized by certain grasses not dominant in the other two.
Lociation	These are subdivisions of the faciations corresponding to differences in soil, exposure, etc.

(Seral stages may also be given names corresponding to the three just preceding; however, the convention is, or was, that they end in *-ies*. Thus, a cattail marsh or a stage in old field succession would be a particular kind of *associes*.)

only one big community that changes according to soil, slope, and other habitat factors. The climax pattern approach considers it more useful and closer to reality to describe this pattern of variation.

Obviously this last approach requires the fewest assumptions. In any case, the first job of the ecologist is to try to understand the workings of the community and only later worry about a classification. Classifications are nevertheless useful for descriptive purposes; the question of whether they can be natural or arbitrary is interwoven with the question of the continuum nature of vegetation (discussed in the next section).

THE INTEGRATED VS. THE INDIVIDUALISTIC COMMUNITY

The *integrated view* of communities is that they are highly organized; the *individualistic view* is that they have little organization. The individualistic view, expounded by H. A. Gleason of the New York Botanical Garden, emphasizes the idea that each individual species tends to be distributed independently of others, occurring where it can disperse and survive. Furthermore, the conditions favoring each species differ. Consequently the organisms of a given piece of ground are simply a gathering of those species finding it within their tolerances.

Proponents of the integrated view hold that the individualistic concept, as usually expressed, gives too little weight to coactions and, especially, their coevolutionary development. Beechdrops, for example, occurs in beech-maple forest because it is able to disperse and survive there; however, one reason for its survival is the presence of beech trees whose roots it parasitizes. The spring ephemerals of the temperate deciduous forest requiring a humus-rich soil for growth, those that grow up and flower early in the spring and die back aboveground by early summer, are adapted to a certain collection of physical factors but this set of factors is produced by the community associates of the spring ephemerals. The integrated viewpoint emphasizes the interactions of members of the community and the effects that the community has on the physical environment. It emphasizes characteristic food chains, close interrelations such as pollination and mimicry, and the influence of certain organisms on soils and microclimates. The integrated view generally leads to a classification of communities. It is believed that, because of the unity developed by community interactions, characteristic combinations of species tend to be repeated and that can be given a name, such as beech-maple forest.

Usually accompanying the individualistic view is the *continuum approach* to describing vegetation.* Communities are supposed to vary continuously in space, with each point of the continuum being equally probable. That which might be called oak-hickory forest and beech-maple forest are simply regions on a continuum, and the combinations between these two are as likely to occur as any others (Fig. 5–8). The continuum is more than one-dimensional. Species might be arranged one way along a moisture gradient and another way along a pH gradient. Their actual

---

*A leading proponent of this approach was J. T. Curtis of the University of Wisconsin.

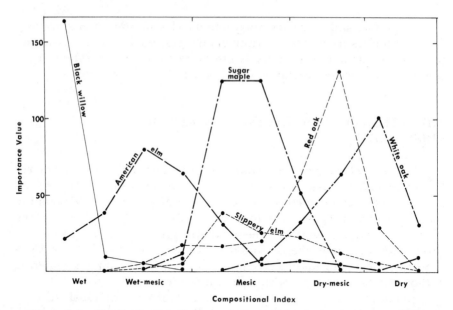

Figure 5–8. The distribution of some major tree species along a continuum of Wisconsin forests from wet to dry. The *importance values* plotted on the y-axis are figures that combine density, frequency, and size, thus measuring three aspects of the "importance" of a species in a forest stand. No two curves have exactly the same shape and location on the continuum, and this remains true when other species like basswood or bur oak are added. (From J. T. Curtis, *The Vegetation of Wisconsin,* Madison, University of Wisconsin Press, 1959, p. 99. Copyright by the Regents of the University of Wisconsin.)

distribution in nature would reflect their response to both, in addition to any other important factors.

There is evidence that continua do exist. Species tend to be distributed in nature according to their tolerances, and the range of one species differs from the range of most other species. It is not clear whether every point along the continuum is equally probable. It seems possible that coactions such as competition and predation, and reactions such as shading and soil formation, make some regions of a continuum improbable. For example, in a region where oak forest and maple forest both occur oak-maple forests may be unusual (Fig. 5–9). It seems possible that maple is able to invade an oak forest only very slowly but, when it does invade, succession to a maple-dominated forest occurs very rapidly. Consequently forests of intermediate composition are rare at any one time.

There is enough evidence on both sides of this question to indicate that the individualistic/continuum and integrated/classification views are not alternatives, one of which must be wrong, but rather complementary, equally valid ways of viewing ecosystems.

STABILITY

Knowing what makes a population or a community stable is obviously of great practical importance. People would prefer their crops and

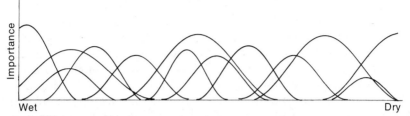

If certain parts of the continuum represented above are probable and others are improbable, then a survey of the land of the region will find many areas having combinations of plants and animals corresponding to the "probable" sections and few corresponding to the "improbable" sections. This situation is represented below, where each mark is a tally for an area having a composition corresponding to a certain point on the continuum. Here there are three clusters of marks corresponding to areas having similar combinations of organisms.

In the example, then, there are three "probable" sections of the continuum. The sections between are improbable; rarely is a piece of land found that has a combination of organisms like those parts of the continuum. Most of the region is occupied by communities like those in the three parts of the continuum enclosed by dashed lines below. The three communities might be, for example, floodplain forest, beech-maple forest, and oak forest.

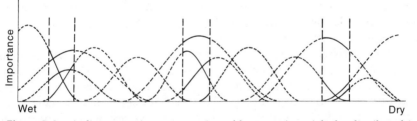

Figure 5–9. A diagrammatic representation of how species might be distributed individualistically, or in a continuum but separate communities might exist because some parts of the continuum are improbable. The probability or improbability of various combinations would result from coactions and reactions of the dominants.

game to be stable at high levels and pests and diseases to be stable at low levels. Unfortunately the question, "Is this population or community stable?" has turned out to be complex or, really, to be several questions that may not be very closely related.

In one sense climax communities are stable and successional communities are unstable; given a reasonably stable environment, the climax community persists while successional communities disappear or change. Of course there are examples of very persistent subclimax communities, and it is clear that climax communities themselves may slowly change but, to a practical individual, the distinction between stable climax communities and changeable successional communities is valuable.

What happens to a climax community if there is a disturbance from the outside—for example, a temporarily changed climate? If stability implies an unchangeable nature, the community should retain the same species in the same numbers, despite drought or flood. Usually, this does not happen, although if the climatic change is short compared with the lives of the adult dominants (such as trees in a forest) there may be little evident change. The response to change usually seems to be more like that described for prairie (see p. 176). In a period of drought some prairie species decreased and others increased. With an end to the drought, the community returned to a predrought structure and composition. Timber cutting in a climax forest produces a similar response. Some shade-intolerant species increase temporarily or invade, but later trends are toward restoring prelogging conditions.

On the community level, then, there seems to be little tendency to resist change. Rather, the community undergoes internal shifts in response to external change, which may maintain some stability (although not constancy) of such traits as cover and productivity. Climax communities also show stability in the sense that they tend to shift back toward their earlier state once the external disturbance has ceased; this is generally expressed by saying that they recover from *perturbations*.

What are the evolutionary advantages of stability of individual species? Very wide swings in numbers will be selected against in the sense that a species which does fluctuate widely may go to extinction, so that only species having those traits which tend to moderate such swings (at least on the down side) continue to be with us. Beyond this, the advantages of population stability to individuals, as affecting their life span and their reproduction, should be considered.

When this is done it seems clear that stability, in the sense of constancy, could rarely be selected for. Regulation of the size of local populations (see Chap. 3) could be selected because mechanisms for recognizing and preventing overcrowding would enhance survival and reproduction. But crowding is relative to resources, at least over a very broad range of densities. Consequently, if conditions become unusually favorable for a species, any genotypes of that species unable to respond to the increased favorability by increasing in numbers will be selected against. We would anticipate that ordinarily species will respond to a better environment by increasing in numbers and, when actual cases are considered, this is usually what we see.

Although the stability of the community is based on the individual organisms composing it, it is worth reemphasizing that some community traits may show more stability than those shown by individual species. If half the species increase and half decline, the net effect may be that the community shows a stable productivity.

Constancy, as useful as it might be to man in his use of ecosystems, seems to be a kind of accident resulting when a species has an effectively constant environment. Species which fluctuate in numbers show no less stability, in any important ecological sense, than ones which do not, as long as the fluctuations tend to be up when the population is low and down when it is high. A little thought will make it clear that this restriction is no restriction at all, since it must apply to every species that does not become extinct.

Recent interest in stability has stemmed from a small book by the English animal ecologist Charles Elton, *The Ecology of Invasions by Animals and Plants*.* Elton summarized several lines of evidence that suggested to him that stability was greater in more diverse habitats or ecosystems. He, and later followers, pointed out, for example, that the predator-prey models of Lotka and Volterra predicted oscillations in numbers, and experiments seem to confirm that oscillations and even extinction are usual results in very simple systems. Furthermore, simple natural systems, such as the tundra, do show wide fluctuations in animal numbers, such as with the snowshoe hare and the lynx. Outbreaks of pest species are most common in simple ecosystems, especially in the very simple monoculture systems of crops favored by modern agriculture. On the other hand, such outbreaks seem very uncommon in the complex natural ecosystems of the tropics. There is also the fact that a high proportion of the species which have become extinct in historical times have been species of the simplified ecosystems of islands.

One basis for the greater stability in more diverse ecosystems is that some perturbation, perhaps an unusually heavy production of food by one plant species, would not be passed along as simply a great increase in one herbivore followed by a great increase in one carnivore, as might occur in a very simple system. Instead, the food would be spread among several herbivore species, each of which was fed upon by several carnivores, so that the large bulge at the plant level would be dispersed into many scarcely detectable bulges at the carnivore level.

A direct connection between diversity and stability was attractive for several reasons, one being that it provided a practical argument against the extreme simplification of ecosystems favored by recent technology. Unfortunately the connection has proved difficult to confirm. For one thing, some of the evidence was probably misinterpreted. One example already mentioned is that it now seems clear that oscillations and fluctuations are not necessarily indications of instability.

Observation has sometimes confirmed that diverse communities are more stable than less diverse ones but there are very simple communities that are apparently very stable, such as those of hot springs. At the same time there is some evidence that species of insects with broad diets may fluctuate more in numbers than those with narrow ones.

Sophisticated mathematical analyses have suggested that, for systems in general, complexity tends to generate instability, confirming everyday experience that simple devices like sundials break down less often than complicated ones like self-winding calendar watches. Of course, mathe-

---

*London, Methuen, 1958.

matical modeling does not tell us that there cannot be a connection between diversity and stability in real ecosystems which, because of evolutionary processes, may not correspond to expectations for "systems in general."

It seems clear that a connection between diversity and stability exists, at least to the extent that extremely simple systems, simplified beyond any natural ecosystem, are unstable. Laboratory cultures of predator and prey usually end with the extinction of one or the other until spatial heterogeneity is added. There is a great deal of observational evidence and several experiments indicating that crop monocultures are more prone to outbreaks of pests than natural communities. The reasons for the outbreaks seem to be related to (1) the high density of the host plant, allowing rapid spread and exponential growth of the specialized pest species; and (2) the absence from the system of such complications as habitat patchiness and alternate hosts, which would maintain satisfactory populations of competitors and predators.

Outbreaks and extinction are the two types of instability that seem to have ecological significance as well as practical importance to man. Some evidence seems to point toward the possibility that these may be discouraged—and stability thereby favored—by spatial diversity. In promoting stability, complexity within the community may be no more important, and possibly less, than complexity in the landscape—the occurrence of patches of several communities. Beech-maple forest is probably more stable than cornfield, but each may be more stable if it exists close to the other and to other community types as well.

LANDSCAPE AND THE BIOME SYSTEM

Ecosystems tend to have a certain look based on broad features of the structure of the vegetation, as well as on the shape and features of the land—the dirt and rocks. In this sense the term "landscape" will be used here. There are at least four main landscape types—forest, savanna, grassland, and desert—with many further subdivisions and additional categories depending on the keenness of your eye.

Forest is a landscape dominated by trees close enough together for their crowns to touch. *Savanna* has trees but these are scattered in a grassy or shrubby area. In *grassland* trees are scarce and grass and other herbs dominate the landscape. In *desert* plants are sparse and often scrubby; here the sand and rock may be more important in the landscape than the vegetation.

Climate is important in producing landscape, particularly through its action in determining the vegetation type. Notice that the relationship with climate is a community relationship. There are grasses, shrubs, and trees, xerophytes, mesophytes, and hydrophytes in almost every climate.

What is important in landscape is the way in which they combine to form communities.

Most systems relating vegetation and climate would put southern Michigan in some moist forest category but we have seen (Fig. 5–7) that the first white settlers found, in addition to moist forest, dry and wet, and also native grassland, bogs similar to tundra, and sand areas similar to desert. Today at least patches of these occur along with still other man-produced landscape types, emphasizing the fact that other factors in addition to temperature and precipitation may be important, at least locally, in determining vegetation type. These other factors include fire and special features of soil and drainage.

Among the more motile animals such as birds and mammals landscape seems to be one of the main bases for habitat selection. In the eastern U.S. a savanna, or forest-edge, landscape can be formed in many different ways and a variety of different plants can be involved. There can be a wholly man-produced community of exotic trees set in mowed lawn with imported shrubbery, or an old field area with goldenrods, blackberries, and cherry trees, or a bog with tamarack trees, blueberries, and sphagnum moss. Many of the birds in these three situations—such as song sparrows, catbirds, and robins—are the same, indicating that their occurrence is related to landscape rather than to plant species.

BIOMES

Climax communities are different for different geographical areas. In 1939 F. E. Clements and V. E. Shelford described the *biome* system for dealing with the geographical distribution of communities. They used as their basic unit the climatically determined landscape types discussed earlier. They defined a biome as a biotic community of geographical extent characterized by distinctiveness in the life forms of the important climax species. The biome, then, is not based on the species of plants and animals present (although detailed study of a biome eventually needs to consider the identity of these plants and animals) but rather it is based, first, on the life form of the most important plants, those that give to the landscape its special character. The biome is viewed as a *biotic* community which, in this sense, means that the plants and animals are considered together as an interacting unit. S. C. Kendeigh has pointed out that although the plant life forms are distinguishing traits of the biomes, the corresponding features for the animals should be special physiological and behavioral traits adapting them to the vegetation and the climate. Seral communities within the region of a biome are considered part of that biome.

The major biomes in North America are temperate deciduous forest, boreal coniferous forest, the broad sclerophyll community of woodland and chaparral, grassland, tundra, and desert (Fig. 5–10). In more tropical

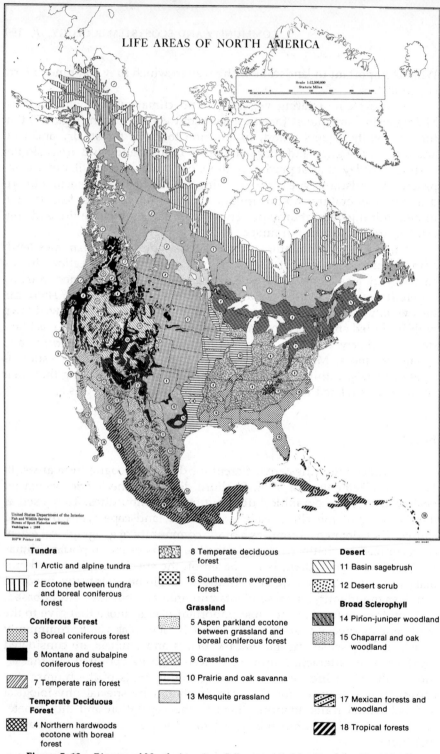

LIFE AREAS OF NORTH AMERICA

Scale 1:12,500,000
Statute Miles

United States Department of the Interior
Fish and Wildlife Service
Bureau of Sport Fisheries and Wildlife
Washington : 1966

BSFW Poster 102

Tundra

☐ 1 Arctic and alpine tundra

▦ 2 Ecotone between tundra
and boreal coniferous
forest

Coniferous Forest

▦ 3 Boreal coniferous forest

■ 6 Montane and subalpine
coniferous forest

▨ 7 Temperate rain forest

**Temperate Deciduous
Forest**

▦ 4 Northern hardwoods
ecotone with boreal
forest

▦ 8 Temperate deciduous
forest

▦ 16 Southeastern evergreen
forest

Grassland

▦ 5 Aspen parkland ecotone
between grassland and
boreal coniferous forest

▦ 9 Grasslands

▤ 10 Prairie and oak savanna

▦ 13 Mesquite grassland

Desert

▧ 11 Basin sagebrush

▦ 12 Desert scrub

Broad Sclerophyll

▦ 14 Piñon-juniper woodland

▦ 15 Chaparral and oak
woodland

▦ 17 Mexican forests and
woodland

▨ 18 Tropical forests

Figure 5–10. Biomes of North America. (Map prepared by Dr. John W. Aldrich, U.S. Bureau of Sport Fisheries and Wildlife Poster 102, Fish and Wildlife Service, U.S. Department of Interior, Washington, D.C., 1966. Terminology slightly modified.)

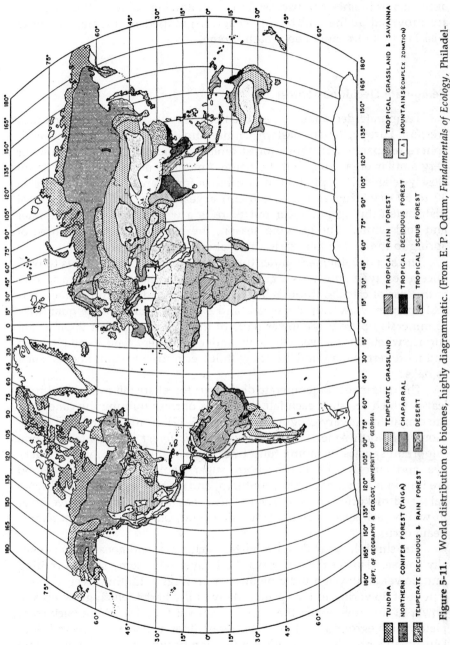

Figure 5-11. World distribution of biomes, highly diagrammatic. (From E. P. Odum, *Fundamentals of Ecology.* Philadelphia, W. B. Saunders, 1971, p. 379.)

regions some additional biomes are present, including tropical savanna, seasonal (deciduous) forest, and rain forest (Fig. 5–11). The main features of the major biomes are discussed in the following sections. Reading lists are provided at the end of this chapter, grouped by biome, for those wishing to know more about each biome.

Temperate Deciduous Forest

Temperate deciduous forest biomes occupy eastern North America, western Europe, Japan, eastern China, and Chile. The northern hemisphere temperate deciduous forest biomes contain many of the same or very similar organisms, but the South American deciduous forest is not closely related. Precipitation in deciduous forest area is fairly high, from 30 to 50 or 60 inches (75 to 150 cm) a year. Rainfall tends to be well distributed through the year but there is a well-defined warm summer and cold winter. The growing season, the period from the last frost of spring to the first of fall, varies from about 140 to 300 days.

The soil development under this vegetation and climate tends to produce a soil type known as a gray-brown podzol. Such soils are acid due to the loss of calcium through leaching. Some deciduous forest trees, notably maples, flowering dogwood, and basswood, bring large quantities of minerals up from the lower parts of the soil and deposit them at the surface when their leaves fall. Soils under such trees tend to be less acid and more fertile than under trees with lower nutrient requirements, such as oaks.

The dominants of temperate deciduous forest are tall trees with broad leaves which are shed each fall. The types of trees forming the canopy of the climax forest vary geographically (Fig. 5–12). In the central part of the biome in North America a *mixed mesophytic forest* may contain several important species including tulip tree, sugar maple, beech, yellow buckeye, and various oaks. Further north sugar maple, beech, and basswood are important; in the western, southern, and eastern regions of this biome oaks and hickories are important. There may be a subcanopy tree layer in which such species as flowering dogwood, blue beech, and hophornbeam occur.

The shrub layer with spice bush, witchhazel, and gooseberry is usually sparse. Usually there is a rich herb layer with some species coming into flower all through the growing season (Fig. 5–13); however, there is a peak of flowering in the spring. Many of the herbs which grow up and flower at this time when the trees are not yet fully leafed die back completely aboveground by June. The forest floor which was covered with Dutchman's breeches, trout lily, and spring beauty in May is almost bare in July (Table 5-4). (Other details of the seasons in temperature deciduous are given in the section on seasonal change in Chapter 4.) Almost all the herbs of the temperate deciduous forest are perennials, although the few

Figure 5–12. The great student of American forests E. L. Braun recognized nine subdivisions of eastern deciduous forest, as shown on this map. The tree species making up the climax forests were the main criteria for deciding regional boundaries. (From E. Lucy Braun, *Deciduous Forests of Eastern North America,* Philadelphia, Blakiston, 1950.)

annuals include such well-known plants as blue-eyed mary and touch-me-not.

Net primary production is usually in the range of 4000 to 8000 kcal per square meter per year. About two thirds of this is stored in wood and thus is not immediately available for herbivore consumption. Most of the rest is stored in leaves. Some 5% of the leaf biomass is consumed by herbivores, mostly caterpillars, while the leaves are on the trees. The rest is not consumed until the leaves fall to the ground. In some deciduous forests consumption and decomposition of one year's leaf production is complete in a year or less so that the litter layer has disappeared by the time of the next autumn leaf-fall. This is usually true of forests dominated by sugar maple and basswood, but in oak forests decomposition and consumption of one year's production takes more than a year, so that a thick litter layer tends to develop.

Some characteristic mammals of the temperate deciduous forest biome of eastern North America are the white-footed mouse, raccoon, eastern chipmunk, southern flying squirrel, gray squirrel, gray fox, and bobcat. The last three of these have disappeared or been reduced in numbers in many areas. Reduction of the forest cover to isolated woodlots has increased the amount of edge, allowing fox squirrels to penetrate most forested areas. Competition with this larger species may be involved in the decline of gray squirrels. The same reduction of forest to isolated stands, each smaller than the home range necessary for a family, may be important for the two carnivores, the gray fox and bobcat. Direct human persecution has probably also been significant.

Characteristic breeding birds include (or included) barred and great horned owls, eastern wood pewee, Acadian flycatcher, black-capped chickadee, white-breasted nuthatch, several woodpeckers, wood thrush,

Figure 5–13. Warren Woods, a beech-maple climax forest in spring (A) and summer (C). Spring ephemerals, such as trout lily (B), grow in great numbers before the trees have leafed out but very few herbs grow up and flower in the dense shade of the late spring and summer. Several herbs and shrubs mature fruit in the summer and many of these are fleshy, animal-dispersed fruits. Most are red but these doll's-eyes (D) are white. Either color stands out in the deep shade.

Table 5–4 NUMBER OF HERB STEMS IN SPRING, EARLY SUMMER, AND LATE SUMMER IN A CLIMAX BEECH-MAPLE FOREST

As shown by these original data from Warren Woods in southwestern Michigan, the forest floor in spring is densely covered by herbs. Many of these are spring ephemerals which sprout, make their vegetative growth, and flower before shading is at its heaviest; they then die back for the rest of the year. Most spring ephemerals are perennial but a few (such as false mermaid weed in the table) are actually annuals that compress their life cycle into the spring and exist as seeds for the rest of the year. Wild leek is different from the other perennials in that it makes only its vegetative growth in the spring; weeks after its leaves have died back it sends up bare flower stalks from its bulbs. By mid-June there are only about 9 plants per square meter and by mid-September the floor is practically bare, with only about two stems per square meter. Most species that are numerous in June and September were already present in April. Beechdrops, a root parasite of beech trees, is one exception: it does not sprout and flower until late in the summer. Not all the spring-flowering plants die back completely in the summer; hepatica, phlox, and ginger are some that do not.

| | Number of Stems (per square meter) | | |
Species	Late April	Mid-June	Mid-September
Spring beauty	123	—	—
Yellow trout lily	107	—	—
Squirrel corn	25	—	—
False mermaid weed	19	—	—
Wild leek	9	—	<0.5
Canada mayflower	1	4	<0.5
Smooth yellow violet	3	3	Present
Sweet cicely	<0.5	1	Present
Pennsylvania sedge	—	1	<0.5
New York fern	—	<0.5	—
Beechdrops	—	—	1
Wild ginger	<0.5	—	<0.5
Hepatica	—	—	<0.5
Plantain-leaved sedge	<0.5	<0.5	<0.5
Blue phlox	—	—	<0.5
All species	About 307	About 9	About 2

red-eyed vireo, cerulean warbler, ovenbird, and scarlet (or summer) tanager. The passenger pigeon, now extinct, was a forest species. Except in the south, winter bird populations are low, composed mostly of sedentary species such as chickadees, nuthatches, and woodpeckers. (The niches of breeding birds of temperate deciduous forest are given in Table 4–6.)

Several reptiles and amphibians are characteristic, including the timber rattlesnake, copperhead, black rat snake, box turtle, wood frog, and gray tree frog. Salamanders are numerous, especially in the central part of the biome. Some characteristic species are red-backed, spotted, blue-spotted, Jefferson's, and marble salamanders.

There is a rich invertebrate fauna. Highest densities are in and on the litter where very large numbers of springtails, mites, millipedes, centipedes, snails, sowbugs, and beetles live. Most of these also occur in large numbers in the soil, along with earthworms and nematodes. The foliage invertebrates of the herbs, shrubs, and trees include mosquitoes, flies, moths, butterflies, leafhoppers, thrips, and spiders.

Some adjustments of animals suiting them for life in this vegetation and climate are the use of trees as nests, den sites, food sources, and feeding and song-perches. Special *arboreal* adaptations (for inhabiting trees) include prehensile tails (white-footed mice), adhesive toe pads (tree frogs), skin flaps used in parachuting (flying squirrels) and, of course, wings, which adapt birds and insects for life in the forest very well but which also have many other uses. Voice and hearing tend to be well developed (birds, insects, squirrels, chipmunks), presumably because the branches and leaves restrict long-distance communication by vision. Adjustment to the harsh winter conditions of low temperature, short days, and decreased food supply takes several forms. Most cold-blooded animals and a few mammals, such as woodchucks and jumping mice, hibernate. A high proportion of birds migrate southward, but some are added by migration from more northerly forests, including the coniferous forest areas. The birds present throughout the winter tend to flock together either for feeding efficiency, predator protection, or both.

Some of our most common animal species today are forest-edge species. These occur in fence-rows, brushy fields, parks, and residential areas—in fact, most of the landscape created by man, with the exception of crop fields and central urban areas, is forest-edge. This landscape type and the species inhabiting it were probably much less common in primeval times, occurring only in such situations as forest openings resulting from wind throw, clearings by Indians, and along the edges of rivers and cliffs. Many familiar forest-edge species must once have been rare relative to their present numbers. Examples are white-tailed deer, red fox, fox squirrel, cottontail, bobwhite, mourning dove, house wren, gray catbird, American robin, common grackle, indigo bunting, American goldfinch, song sparrow, and common garter snake.

In the southeastern U.S. much of the land is covered with pine and pine-oak forests and savannas. Where these communities are protected from fire they are replaced by broad-leaved forest, either temperate deciduous or in parts of the Gulf States and Florida broad-leaved evergreen forest of live oaks and magnolia. Lightning-set fires are a common occurrence in this region and several of the plants making up some of the pine communities, such as the longleaf pine, show adaptations to recurrent burning (see Fig. 2–15). Pine forests have probably been a feature of southeastern North America for a very long time. Some characteristic animals of these southeastern pine communities are the golden mouse, southeastern pocket gopher, pine warbler, rufous-sided (white-eyed) towhee, mockingbird, Carolina wren, red-cockaded woodpecker, brown-headed nuthatch, and (formerly) Carolina parakeet.

In the Time-Life book *The World We Live In** there is a good chapter on forests in which temperate deciduous forests are referred to as "the

---

*New York, 1955.

woods of home." The phrase has a nice ring to it and a certain amount of truth but it is incorrect in several ways. Temperate deciduous forest biomes are home to a great many people in western Europe, eastern Asia, and eastern North America but other biomes are home to such people as Eskimos, Arabs, southern Europeans, and Polynesians. Even to those of us in North America, temperate deciduous forest is rarely home in the sense that we live in one. Forest edge, or temperate savanna, seems to be the landscape we prefer. Where our ancestors settled in forest they cut openings in it; when eventually they moved out onto the prairie they planted trees. The Indians were similarly uninterested in climax deciduous forest as a place to live. Through the summer it was too buggy, and the nuts and berries gathered by the squaws and the game that the braves hunted were found in forest edge and open oak forest (Table 5-5).

Table 5–5 SOME IMPORTANT FOOD PLANTS OF INDIANS IN THE UPPER GREAT LAKES REGION BY HABITAT*

R. A.Yarnell has listed about 120 plants used by Indians at the time they first came into contact with Europeans. Although virtually every habitat contained some food plants, the partial lists in Table 5–5 indicate correctly that a greater number of species and more important species were in marsh and forest edge. The climax beech-maple forests had relatively few important species; the same was true of coniferous forests.

Habitat	Food Plant
Beech-maple forest	Sugar maple, wild leek, spring beauty, basswood, toothwort, prickly gooseberry, beech, paw-paw
Oak and oak-hickory forest	Bracken fern, dryland blueberry, shagbark hickory, white oak, black oak
Swamp and bog forest	Silver maple, yellow birch, jack-in-the-pulpit, black huckleberry, wild currant, black walnut, skunk cabbage, hackberry
Coniferous forest	White pine, creeping snowberry, bunchberry
Marsh, bog, and lake	Great bulrush, American lotus, yellow pond lily, marsh marigold, fragrant pond lily, arrowhead, wild rice, cranberry, highbush blueberry, cinnamon fern
Prairie and bur oak savanna	Butterfly weed, wild strawberry, hazelnut, bur oak, wild pea, dewberry
Forest edge	Common milkweed, common elder, raspberry, blackberry, chokecherry, black cherry, plum, nannyberry, wild grape, hawthorn, bittersweet

*Species list from Yarnell, R.A., "Aboriginal relationships between culture and plant life in the Upper Great Lakes region," Ann Arbor, University of Michigan Museum of Anthropology, Anthropological Paper No. 23, 1964.

Temperate Grassland

The climate of regions occupied by grassland is variable. Summers are usually hot but winters may be cold to mild. Rainfall varies from low to moderate but there is usually a dry period in late summer, fall, or winter. Spring or fall fires are characteristic. The soils which develop under the grass cover in these climates are prairie soils and chernozems which are typically fertile, high in organic matter, and nearly neutral to basic.

Grasslands occur geographically in the interior of continents in central North America (where the eastern, tall grass portion was called *prairie* and the western, short grass portion *plains*), eastern Europe (in Hungary, *puszta*), central and western Asia (in Russia, *steppe*, a term also generally used to refer to any short grass grassland), Argentina (*pampas*), and New Zealand.

The vegetation consists of perennial grasses and herbs (*forbs*), particularly legumes and composites. Height, denseness, and productivity vary with rainfall. Where rainfall is high the grass may grow to 2 meters and form a thick sod; in the more arid regions plants are low and sparse with no continuous sod. Net annual production ranges between 800 (dry) and 6000 kcal/m². Light to moderate grazing may stimulate grass production. Storage is mainly in roots and underground stems but much of each year's production is in leaves and roots, which die and become available immediately to consumers and decomposers. Prominent year to year variations in plant cover and production based on variations in rainfall also occur. Flowering occurs throughout the growing season with the flowers usually borne just above the level of the growing foliage. There is a peak of flowering in late summer.

Some characteristic plants of the North American grasslands include *tall grasses*—big bluestem, Indian grass, slough grass; *mid grasses*—needle grass, dropseed, little bluestem, side-oats grama, June grass; *short grasses*—buffalo grass, blue grama; *forbs*—many species of rosinweeds, goldenrods, asters, mints, and legumes.

Some characteristic animals of North America include jackrabbits, prairie dog, 13-lined ground squirrel, pocket mice, coyote, badger, black-footed ferret, pronghorn, bison (Fig. 5–14), prairie chicken, burrowing owl, horned lark, bobolink, meadowlarks, Savannah sparrow, and longspurs. The reptile fauna is dominated by snakes such as the plains garter snake, bullsnake, massasauga, and western rattlesnake. Insects are numerous, especially grasshoppers.

Some of the characteristic adjustments that animals show to life in this biome include adaptations for grazing (for example, specialized teeth) and for rapid travel. Herding is common in mammals and flocking in birds in winter. Burrowing and the use of burrows is common, extending even to birds (burrowing owl). Camouflaging brown streaked coloration

Figure 5–14. More than any other animal, the bison is a symbol of the native North American grassland—and its disappearance. Ranging from the small prairies of the East to the Rockies, bison numbered 60 to 75 million in the eighteenth century, when horses brought by the Spaniards became widespread among the Plains Indians. The Indians concentrated their hunting—and, with success, most aspects of their life—on the bison herds. It was a rich, glorious time. Then the white settlers began to arrive, and the railroads, and the cavalry. One tactic in the subjugation of the Indians was to remove their food base by the extermination of the bison. Every year through the early 1870's millions of bison were killed; by 1878 the southern herd was gone, by 1883 the last of the Dakota bison had disappeared, and by 1889 there were only an estimated 150 animals alive in the wild in the entire country. There are more than that now, scattered about on wildlife refuges. The animals in the picture are in Montana. (Photograph by H. W. Henshaw, courtesy U.S. Fish and Wildlife Service.)

is prevalent among birds. Many birds have flight songs in this habitat where elevated singing perches are scarce. Hibernation and migration are well developed in the more northern grasslands; dry season aestivation also occurs.

The moister grasslands have been converted into the earth's richest agricultural lands. Many small prairies occurred as patches in the forest cover at the eastern end of the prairie peninsula. They were usually surrounded by oak savanna called "oak openings" by the settlers.

Tundra

At high latitudes beyond the tree line in northern Eurasia, North America, and Greenland where tundra occurs, the growing season is short, 60 days or less, and winters are very cold. Surprisingly, precipitation is low (usually less than 25 cm per year) and most of it falls as rain in summer and autumn. There are great changes in day length from summer to winter.

Tundra soils remain frozen in their lower layers throughout the year (*permafrost*). Drainage is poor and soils are often waterlogged in summer. Ponds, lakes, and bogs (*muskegs*) are numerous. Vegetation is low and mat-like, with grasses, sedges, mosses, and lichens dominating. Many of the herbs have large bright flowers. There are almost no annuals. Woody plants are mostly chamaephytes such as dwarf willows and birches, which are only several inches in height. Some other characteristic North American plants (the tundra flora is not a diverse one) are bilberry, cottongrass, avens, arctic poppy, and reindeer moss.

Net annual primary production is low, about 800 kcal/m², with much of it stored in bulbs and roots. Daily rates of production during the short growing season are about as high as in temperate regions.

Animal adjustments in the tundra include flight songs and white coloration, especially in winter. There is little hibernation since few refuges from freezing are available. Small mammals stay active under the snow. Migration is common among birds which may pair before arriving for their brief stay on their breeding grounds. The barren-ground caribou migrates to wintering grounds at the northern edge of the boreal forest.

Expectably, there are few reptiles and amphibians in tundra. Varieties of insects are few but numbers of certain species such as mosquitoes, black flies, and deerflies may be enormous. Characteristic birds and mammals include rough-legged hawk, snowy owl, jaegers (Fig. 5–15), Lapland longspur, snow bunting, many species of waterfowl and shorebirds, arctic fox, lemmings, arctic hare, barren-ground caribou, and musk-ox.

Western technology has been slow in reaching the tundra. The cold, the six-month-long winter nights and the problems of constructing roads

Figure 5–15. Long-tailed jaeger in Alaska. Although related to the gulls, jaegers are predators and, like most predators in tundra, they eat lemmings. One of the characteristic features of tundra is the oscillatory nature of lemming populations. When lemmings are scare some of the predatory birds and mammals may not breed at all; when lemmings are abundant they breed in large numbers and may even have larger litters (or clutches). (Photograph by L. H. Walkinshaw.)

and buildings on the unstable permafrost-based soils have acted as barriers. As late as 1940 the population of Alaska was less than 75,000. Today it is five times larger and growing at the rate of 3% per year. The discovery of oil on the North Slope in 1968 led to one of the major environmental battles of the early 1970's over the construction of a trans-Alaska pipeline across 800 miles of tundra and boreal forest. Construction of the pipeline was finally authorized in 1973, with the addition of several environmental safeguards. How effective the safeguards are—how severe the melting of permafrost from the hot oil flowing in the pipe will be, what the soil and vegetation effects are, how badly animal activities such as caribou migrations will be disrupted—remains to be seen. But it is clear that the trans-Alaska pipeline is only the first of a succession of drastic changes in the tundra as western man and his technology invades this fragile ecosystem.

Above the timberline on mountains throughout the world occurs vegetation generally referred to as *alpine tundra*. The largest area is on the Tibetan Plateau; in North America, alpine tundra occurs at higher elevations in the western mountains as far south as Mexico but only into New England in the Appalachians. The species of alpine tundra are mostly different from those of arctic tundra; furthermore, the alpine tundra areas of the various continents share few or no species. As compared to arctic tundra alpine tundra has more precipitation, less extreme photoperiods, low oxygen concentrations, and high ultraviolet radiation. Conditions are probably less severe in winter but, with cold nights and high winds, more severe in summer. Alpine soils lack permafrost (except in the far north) but are thin over bedrock. The vegetation has more grass and sedge and less shrub and lichen than in arctic tundra. Characteristic animals include marmots, mountain goat, Dall's or mountain sheep, pikas, white-tailed ptarmigan, and rosy finches. Migration either southward or to lower elevations (called *altitudinal migration*) is prevalent, as is white coloration for animals which remain for the winter.

Coniferous Forest

Coniferous forest occurs as a broad band across northern North America and northern Eurasia (called *taiga* in Russia) and also extends southward at higher elevations in the mountains. The climate within this biome is characterized by cool summers, cold winters, and a growing season of less than 130 days. Precipitation is between 40 and 100 cm per year, with heavy snow in winter.

The soils which develop under the needle- or scale-leaved evergreen forests that dominate the region are podzols, acid and infertile, with thick litter. Peat and muck areas are frequent. In mountainous areas soils may be shallow and rocky.

Often one or two species, especially white spruce and balsam fir, dominate large areas. Red spruce replaces white spruce in the Appalachians. Black spruce often dominates boggy areas but may be replaced by white cedar. Pines and also broad-leaved trees such as balsam poplar, quaking aspen, and birches often dominate seral communities. Shrubs often have fleshy fruits; examples are blueberries, gooseberries, and dwarf cornel (dogwood). Characteristic herbs include starflower, twinflower, and sarsparillas; however, the herb layer is usually sparse in the heavily shaded conifer forests.

Despite the short period between spring and fall frosts, productivity may be high because the evergreen habit allows photosynthesis on any day when conditions are even temporarily favorable. Net annual production may be as high as 8000 kcal/m² but more often is between 2000 and 3000 kcal/m².

Characteristic animals in North America include snowshoe hare, red squirrel, porcupine, gray wolf, red-backed mouse, bog lemming (grassy or sedge bogs), marten, lynx, moose. Birds include goshawk, grouse, three-toed woodpeckers, olive-sided flycatcher, gray jay, red-breasted nuthatch, boreal chickadee, winter wren, Swainson's thrush, kinglets, many species of wood warblers (Fig. 5–16), purple finch, pine grosbeak, and white-throated sparrow. Breeding populations of birds tend to be very high but most migrate out in winter. There are few reptiles; toads, mink frog, and wood frog are about the only amphibians.

Arboreal habits are well developed. Adaptations for travel in deep snow are seen in several animals such as the lynx and the willow grouse. Birds often have high pitched songs; many have adaptations such as crossed bills for feeding on conifer seeds.

At the northern border of the boreal coniferous forest is an extensive ecotone with tundra. Some species probably more characteristic of the ecotone than of either tundra or forest are rough-legged hawk, hawk-owl,

Figure 5–16. The most abundant birds of the eastern boreal forests are the wood warblers. Here a myrtle, or yellow-rumped, warbler feeds three newly fledged young. (Photograph by L.H. Walkinshaw.)

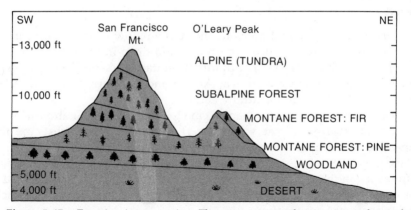

Figure 5–17. Zonation in mountains. The average annual temperature drops about 0.6°C for each 100 m of elevation (or about 1°F for every 325 feet). This, along with other differences in climate, is why mountains have zones of vegetation that resemble the communities of more northerly, colder regions. San Francisco Mountain is on a plateau rising up from the Painted Desert (part of the Sonoran desert) of northern Arizona. A piñon-juniper woodland zone occurs at lower elevations on the plateau. Above this is montane forest, of ponderosa pine, and above it, montane forest of Douglar fir. At about 2800 m or 9200 ft subalpine forest of Engelmann's spruce begins. Timberline, with stunted and prostrate spruce and foxtail pine, is at around 11,000 feet, and alpine tundra of saxifrages and sedges lies above this. The elevations of the different zones are affected by slope exposure, beginning lower on warm, south slopes and higher on cold, north slopes. (From C. Hart Merriam, "Results of a biological survey of the San Francisco Mountain region and the desert of the Little Colorado, Arizona," *North American Fauna*, No. 3, Plate 1, 1890.)

great gray owl, common redpoll, tree sparrow, and white-crowned sparrow.

Montane and Subalpine Forests. These forests of the mountains of the western U.S. (Fig. 5–17) are sometimes considered as a biome separate from the boreal (northern) coniferous forest. They contain several species of conifers not found elsewhere. Among the animals, there are fewer wood warblers and several distinctive species such as the grizzly bear, Steller's jay, the dipper (along streams), and Townsend's solitaire. Bird populations are lower than in the northeastern parts of the biome.

Temperate Rain Forests. Especially distinctive are these coastal coniferous forests. As we shall see in a later section, their relationships in geological development are closer to temperate deciduous forest than to boreal forest. They grow in a climate of moderate temperatures with heavy winter rainfall and almost daily fog in the summer. Cape Disappointment, Washington, averages over 2500 hours of heavy fog per year. These forests are among the tallest in the world, with heights of 60 to 90 meters. Fallen trees and a thick layer of litter on the soil support luxuriant growths of mosses and ferns. Western hemlock, western red cedar, Douglas fir, Sitka spruce, and redwood are important tree species. The wapiti and the chestnut-backed chickadee are among the characteristic animal species. Southern hemisphere temperate rain forests in Chile, Tasmania, New Zealand, and Australia share few, if any, plant and animal species with the North American forests.

Desert

Desert occurs in arid areas in the rain shadow of mountain ranges, along coasts next to cold ocean currents, and deep in the interior of continents. Extensive deserts occur in intermountain and southwestern North America, western South America, Arabia and northern Africa, southern Africa, central Asia, and central Australia.

Rainfall is typically less than 25 cm (10 inches) per year and the evaporation rate is high. In the warm desert of the Southwest frosts are rare and there is a long growing season. In the cold desert of the Great Basin, frosts may occur from September to June. In both, large day-to-night differences in temperature are usual. Desert soils (gray in cold and red in warm) are high in inorganic nutrients and low in organic matter. Rocky and saline areas are frequent.

Vegetation of the desert is characterized by sparse xerophytes. Shrubs, usually with small, deciduous leaves, are prominent. Many plants are spiny. Annuals grow and flower after rains (Table 4-6).

Characteristic plants of the cold desert (basin sagebrush) in North America include shadscale, sagebrush, rabbitbrush, winter-fat, Indian rice grass, and chichalote. Cactuses are not common. In the warm desert (desert shrub) creosote bush is widespread. Subdivisions of the warm desert sometimes recognized are the *Mohave* desert, with Joshua trees and other yuccas, Mohave sage, spring senna, and Parry saltbush (Fig. 5–18). The *Sonora* desert (the most diverse portion) has smoke tree, sages, crucifixion thorn, the familiar saguaro cactus, ocotillo, white bur sage, and a large variety of ephemerals. The *Chihuahuan* desert has yuccas, agaves, many small cactuses, sotol, and tarbush. Floodplain vegetation with willows and cottonwood occurs throughout the desert, as does marsh (tule) vegetation of bulrushes, cattails, and saltgrass.

Net primary production is directly related to rainfall, from 0 to about 200 kcal/m^2 per year. Much of each year's production is stored in wood or seeds.

Large mammals are scarce in the desert although the pronghorn formerly occurred in the cold desert. Occurring widely throughout the desert are the black-tailed jackrabbit, white-tailed antelope squirrel, grasshopper mice, pocket mice, kangaroo rats, wood rats, coyote, kit fox, western spotted skunk, loggerhead shrike, black-throated sparrow, side-blotched uta (a lizard), horned toad, and whip-tailed lizard. In the cold desert are sage grouse, sage thrasher, sage sparrow, Brewer's sparrow, and numerous lizards such as the sagebrush lizard. In the warm desert are the cactus mouse, collared peccary, Gambel's quail, roadrunner, elf owl, lesser nighthawk, gila woodpecker, vermilion flycatcher, verdin, several thrashers, and phainopepla, desert tortoise, many lizards, gila monster, rosy boa, and sidewinder.

As in grassland, the use of burrows is widespread. Many organisms are nocturnal and there is little midday activity, even among diurnal

Figure 5–18. The Joshua tree is the most characteristic plant of the Mohave desert. It is a yucca and, like others, its flowers are pollinated by yucca moths. At 8 to 10 meters it is the tallest plant of the community, and it provides a nesting site for birds such as the Scott's oriole. Photo by P. S. Bieler in Nevada National Forest, courtesy U.S. Forest Service.

forms. Pale coloration, matching the pale desert floor, is frequent. Some animals show adaptations to life and locomotion in loose sand, such as sidewinding. Aestivation during the dry season occurs in many invertebrates and some vertebrates. Breeding seasons tend to occur after rain. Granivory (seed-eating) is well developed.

The various desert areas are new geologically, and the deserts of the different continents are dominated by unrelated species. In Africa, for example, there are succulent euphorbias that look much like the cactuses of North America.

Succession occurs slowly, if at all, in much of the desert, because the community's reactions on the physical environment are slight. Areas studied by botanists early in this century were much the same, often with the same individual plants, when restudied 30 years later. Recently there has been a great deal of disturbance by off-the-road vehicles, cactus collectors, and developers. Recovery will be slow.

"Making the desert bloom"—irrigating to make it productive agriculturally—is a recurrent theme in plans to increase the world's food production. Of course the desert blooms now, but with cactuses, ghost flowers, and Panamint daisies. Making it bloom with crops entails several problems. Getting enough water to the desert without an impossible expenditure of energy is one difficulty. The deposition of salts in croplands as irrigation water evaporates is another problem, one which has already led to the abandonment of some irrigated farms in the American West.

Broad Sclerophyll

These shrublands and woodlands occur in a "Mediterranean" climate with mild, damp winters and hot, dry summers. In North America broad sclerophyll vegetation occurs in the West, especially in California and Mexico and east to New Mexico. Elsewhere in the world it occurs around the Mediterranean (*maquis* or *garigue*), on the southern coast of Australia (*mallee scrub*), at the southern tip of Africa (*cape scrub* or *fynbos*), and in central Chile.

In North America, the biome varies from dense scrub (*chaparral*) to open stands of trees (*woodland*). The dominant life form is a shrub or small tree with thick, leathery, evergreen leaves (*sclerophyll* means "hard leaf"). Most sprout from their roots readily after fires. (The role of fire in chaparral is discussed in Chapter 2.) Grasses and forbs may be abundant in the less dense woodlands and in chaparral when it has been opened up by fire. The forbs often have showy, fragrant flowers.

Soils vary from ones similar to those of deserts to podzols in the most forest-like situations. Red clays called *terra rossa* are characteristic of the Mediterranean area.

Characteristic plants in North America include chamise, several species of oaks (such as canyon live oak, blue oak, Emery oak), manzanita, mountain lilac, and coffeeberry. Net annual production probably averages around 3000 kcal/m², growing season moisture being a limiting factor. In chaparral much of each year's production is tied up in litter, with the energy later lost in fires.

Characteristic mammals in North America include the grizzly bear, mule deer, bobcat, and brush rabbit (Fig. 5–19). Some birds particularly characteristic of chaparral are California quail, Anna's hummingbird, wrentit, California thrasher, gray vireo, and rufous-crowned sparrow. Woodland species include Nuttall's woodpecker, bridled titmouse, Hutton's vireo, and black-throated gray warbler. For birds, short wings, a long tail, and dull (especially brown) colors are frequent. There is some migration out by larger mammals in the dry season. Amphibians are not

Figure 5–19. Mule deer are characteristic large mammals of woodland and chaparral. In some regions they occupy these communities throughout the year; elsewhere they spend the summer at higher elevations in coniferous forest and come down for the winter. These deer were photographed in the Malheur Refuge, Oregon, by R. C. Erickson. Photo courtesy U.S. Fish and Wildlife Service.

common. The reptiles occurring in these vegetation types are mainly those more characteristic of other vegetation.

Certain types of woodland contain conifers with or without oaks as codominants. Sometimes regarded as a separate biome is **piñon-juniper woodland** with piñon pines and several species of junipers. It occurs mostly as a narrow belt in mountains from Wyoming to Mexico. Characteristic animal species often show adaptations to using pine seeds and acorns. Some characteristic birds are the plain titmouse, piñon jay, and scrub jay.

In the Mediterranean region this climate and general vegetation type was the important early center of civilization. (In the U.S. for the past 50 years it has attracted an increasing percentage of the population; however, its status as a center of civilization is not yet clear.) From the descriptions by classical writers of the "thick woods and gigantic trees" of southeastern Spain, of "wooded Samothrace," and similar allusions, it is evident that parts of the Mediterranean region were more heavily wooded two to three thousand years ago. Timber cutting for shipbuilding and other purposes, clearing for cultivation (followed by abandonment as Rome declined), fires, goats, soil erosion, and climatic changes may have been involved in the conversion of such forests to the scrub that is climax over much of this region.

Tropical Rain Forest

Located near the equator in central America, the Amazon and Orinoco basins of South America, western Africa, and the Indonesian region, tropical rain forest occurs in a climate characterized by high temperatures and high rainfall, usually 200 cm (80 inches) or more per year. There is little seasonal variation in temperature or in day length. Reddish or yellowish lateritic soils are characteristic.

The vegetation consists of broad-leaved evergreen trees in three layers. The forest is very open beneath the canopy due to the sparseness of herbs and shrubs; however, vines and epiphytes are abundant. ("Jungle" is dense vegetation resulting from secondary succession.) Trees often have buttress roots; leaves of many plants have drip tips (Fig. 5–20). Decay is rapid and little litter builds up. Cycling of nutrients tends to be direct from plant to plant via digestion of dead plant material by mycorrhizae (fungi associated with tree and other plant roots); thus, no large nutrient reservoir exists in soil or litter. Both flora and fauna are very rich. For example, about 2500 species of trees are known just from the Malay Peninsula, as compared with about 800 for all of North America north of Mexico. Every hectare contains many species; the dominance shown by one or a few species as in temperate oak or maple forests is not seen. Some important plant families are Annonaceae, Araceae, Bombaceae, Dipterocarpaceae, Ebenaceae, Lauraceae, Leguminosae, and Meliaceae. Orchids and bromeliads are important epiphytes.

Net annual production is very high, probably regularly over 8000 kcal/m²; however, a high percentage is stored in wood.

Important animal groups include monkeys, sloths, bats, lizards,

Figure 5–20. Buttress roots are a feature of many tropical forest trees. Here the base of a large fig is seen in a cleared area in tropical dry forest at Tikal, Guatemala. There are several hypotheses as to the function of the buttress roots; the most satisfactory seems to be that, for these tall trees lacking tap roots, the buttresses serve to reduce stresses from the wind on the shallow root system. The leaves along the trunk are herbaceous vines. (Photograph by Richard W. Pippen.)

snakes, hummingbirds, toucans, and parrots. Many species, low populations, and extreme specialization are characteristic. Arboreal adaptations, such as prehensile tails and adhesive disks, are prominent. Bird nests and insect cocoons are often suspended from trees. Voice and hearing are well developed. Specialization to nectar and fruit feeding is much more common than in temperate biomes. Flocking is well developed in birds.

Where there is a pronounced dry season, tropical seasonal forests (such as monsoon forest) replace tropical rain forest. In these, some or most of the trees lose their leaves during the dry season. There are fewer species and stratification is less complex.

Rain forest and other natural ecosystems of the tropics are being destroyed at an almost unbelievable pace. It has been estimated that 120,000 square kilometers of tropical forests are cleared every year. This is an area, every year, larger than the state of Ohio. In 1957 forest still occupied 74% of the Malay Peninsula. By 1971 the figure was 61%, and one biologist has predicted that the forest will be totally destroyed in less than 20 years. Another biologist has given the Amazonian rain forest only a couple of years longer.

Forests are being cut for timber and plywood, much of it exported to the developed nations, for highway building to open up previously inaccessible regions to settlement by the growing population, for agricultural lands to feed the growing numbers, and for metal mining, largely for export. Environmentalists have suggested the possibly serious consequences (to climate, for example) of eliminating the earth's largest source of photosynthesis and adding the carbon contained in its organic matter to the atmosphere. They have suggested that the scrub savanna that will replace much of the forest will not be productive enough to support dense populations of humans, and that soil deterioration will lower production further leading, on some sites, to virtual deserts. The extinction of thousands of species of plants and animals, many of them not yet even named by science, has been mentioned.

The governments of the developing countries are generally unsympathetic to such concerns. Many tropical countries have nature preserve systems; however, they tend to regard suggestions that their nation should, in effect, become one large wildlife sanctuary as being merely another tactic in the attempt by the developed nations to retain domination. And so the rain forests go the way of the North American prairie.

Tropical Savanna

There are three seasons in tropical savanna—warm and rainy, cool and dry, and hot and dry. Rainfall is variable but usually between 90 and 150 cm (30 to 60 inches). Fires are prevalent in the dry season. Soils are variable; many are lateritic but some are similar to those of temperate grassland.

Figure 5–21. A fresh burn in tropical savanna in Belize. The trees in the background are oaks. (Photograph by Richard W. Pippen.)

Much of tropical Africa, large areas of central America and eastern South America, and parts of southeast Asia are occupied by this vegetation type, which consists of grassland with scattered good-sized trees or patches of open forest. The grasses are often tall. Trees, often legumes, tend to be flat-topped and thorny. They may be either deciduous or evergreen (Fig. 5–21).

Net annual production is variable with climate, from 800 to 8000 kcal/m^2. Much of it goes into foliage which is available to herbivores. Large grazers occurring in herds are common, as are large carnivores preying on them. Migration and reproduction tend to be correlated with the rainy season.

Flora and fauna vary from one region to another. In Africa this is the biome of such familiar animals as the zebra, wildebeest (Fig. 3-14), giraffe, lion, cheeta, hyena, and secretary bird. Several African nations have set up wildlife preserves in savanna.

AQUATIC ECOSYSTEMS

LAKES, PONDS, AND MARSHES

The fraction of the earth's surface covered by fresh water is trifling, about 2%. The percentage of the total water of the biosphere that is in lakes and rivers is even less impressive, less than 0.2%. Because of man's interest in inland waters *limnology*, the science of fresh water (physical, chemical, and biological aspects), has received what might seem a disproportionate amount of study. Man is interested in lakes and rivers as scenery, for boating, fishing and related recreation, to dump sewage in and to get water out for drinking, industry, and irrigation. He is also interested in lakes and rivers scientifically as links in the hydrological cycle and as homes of a set of organisms with modes of life very different from those of terrestrial organisms.

This section deals with lakes, ponds, and marshes—or, in other words, areas of standing fresh water. (The classical term used to refer to standing, or still, water is *lentic*.) There is no clear distinction between lakes, ponds, and marshes. Basically, ponds are small, shallow lakes. Marsh refers to a vegetation type, the floating-leaved and emergent vegetation that grows in shallow water which often occurs as a fringe around lakes.

Physical Factors in Lakes. Water as a habitat has a number of features that need to be mentioned. It has a high specific heat; that is, a large amount of heat must be added to raise the temperature of water. This means that daily temperature changes are slight and seasonal changes slow compared with terrestrial habitats.

For most lakes most of the heat input is through solar radiation. Some fraction of the sunlight that hits a lake surface is reflected away. As it passes through the water, the remaining light energy is absorbed and converted to heat. About half of the total light energy is absorbed in the first meter, but some light penetrates great depths. The rate at which light is absorbed by the water varies by wavelength; 95% of the red light is lost in the first 6.5 meters but 95% of the blue light is not lost until a depth of about 550 meters. Light penetration can, of course, be very much reduced in *turbid* waters, those containing suspended materials. The upper zone of the lake in which oxygen production in photosynthesis exceeds the use of oxygen is called the *trophogenic* (or *euphotic*) *zone*. The depth at which oxygen release and oxygen absorption balance is the *compensation depth*; below it lies the *tropholytic zone*. The compensation depth is the level at which light intensity is reduced to about 1% of full sunlight. This depth in feet or meters varies greatly depending on turbidity and water color.

Water is about 775 times as dense as air. This allows it to provide the buoyancy that makes swimming possible but it also means that water provides much greater resistance to movement than does air. Unlike most materials water is densest at a temperature above its freezing point—ice floats—and this is a major feature of water as a habitat. It would be a different world if lakes froze from the bottom up.

The temperature-density relationship of water is important in producing patterns of temperature change with depth, referred to as *thermal*

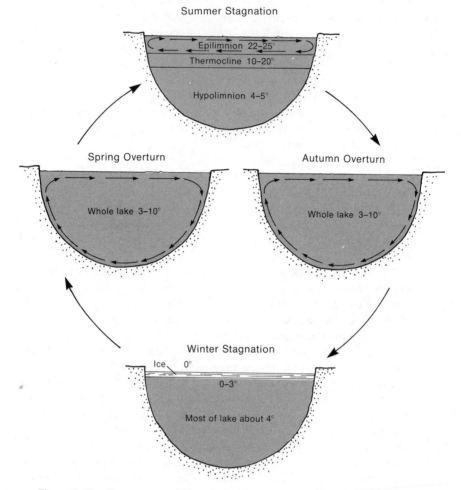

Figure 5–22. Temperatures (°C) with depth for a fairly deep, temperate-zone lake for four seasons. The small arrows show water circulation caused by wind blowing across the lake.

stratification (Fig. 5–22). The pattern for most of the deeper lakes of the temperate regions of the earth is as follows: In the spring, sunshine and warm air temperatures melt the ice cover and bring the upper layers of the lake to the same temperature as the lower, about 4°C. When this occurs winds can cause the whole lake to circulate. During the period of this *spring overturn*, the lake temperature may stay around 4°C or it may rise several degrees higher.

Usually during a warm, calm period the temperature of the surface waters are raised a few degrees higher than the lower portions of the lake. This warmer, lighter water tends to circulate by itself on top of the cooler heavier water. From this develops the typical summer stratification, with an upper *epilimnion* of warmer, circulating water, and a lower *hypolimnion* of cold water that has little circulation. The area between, in which temperature change with depth is rapid, is called either the *metalimnion* or the *thermocline*. The temperature of the hypolimnion may be 4°C but it is often higher and may increase through early summer.

In the autumn the lake loses more heat than it gains. The surface waters cool to the temperature of the upper part of the metalimnion so that, in effect, more and more of the lake is drawn into the epilimnion. Eventually the autumn overturn occurs; the upper waters cool to about the same temperature as the lower waters and the whole lake circulates.

As air temperatures drop through the early winter the lake temperatures also drop but circulation continues. As surface waters cool below 4°C they no longer sink, but the density differences in the range between 0 and 4°C are so slight that even small amounts of wind can keep the lake circulating. Once the lake has frozen, however, winter stratification can occur. Here the lighter, colder water of about 0 to 2°C lies on top of the warmer water of around 4°C.

Lakes with two overturns per year, as just described, are called *dimictic* lakes. In other-than-temperate climates, other patterns exist involving a single overturn, or several, or none.

The concentration of oxygen is lower in water than in air and is also very variable from time to time and from place to place. In general, it is very low oxygen concentrations that are of the times and places of the most ecological interest. One situation in which oxygen may become very low is in ponds or shallow lakes with rooted hydrophytes covering the whole bottom. At the end of the summer, when these die back, their rapid decomposition may lower oxygen levels through the whole lake enough to kill much of the fauna. Die-offs of fish under such circumstances are called "summer kills."

Oxygen depletion in the hypolimnion during the summer is likely to occur if the hypolimnion is below the zone of effective light penetration, thus having little or no photosynthesis occurring in it. In this situation, where the hypolimnion is sealed off from the oxygen of the air and no oxygen is being added by plants, the respiratory activities of the animals and bacteria can exhaust the oxygen content of the hypolimnion by late summer. Oxygen depletion during this *summer stagnation* period is more apt to occur in highly productive lakes where there is more organic material for decay. The third situation in which oxygen may be severely depleted is in winter when there is a heavy snow cover over the ice. Light reduction slows or stops photosynthesis in the algae; if the snow cover lasts several weeks, decomposition and animal respiration can reduce the oxygen supply below the tolerance limits of some organisms with "winter kills" of fish as a result.

Communities and Organisms. For descriptive purposes, three regions of lakes are usually recognized (Fig. 5–23). These are the *littoral*

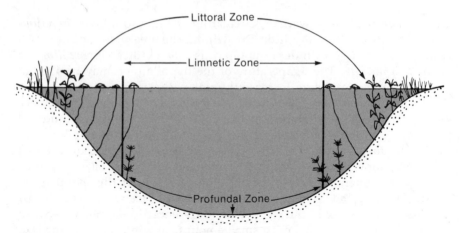

Figure 5–23. Littoral, limnetic, and profundal zones in a lake.

zone, the shallow water around the shore occupied by rooted vegetation, the *limnetic zone*, the open water beyond the littoral zone, and the *profundal zone*, the bottom below the limnetic zone. Sometimes the deeper parts of the open water are included with the profundal zone. In shallow lakes the whole basin may be occupied by rooted vegetation, in which case neither limnetic nor profundal zones exist.

There are basically five life habit types among the organisms of lakes: benthos, periphyton, plankton, nekton, and neuston.

The *benthos* is composed of the bottom-dwelling organisms. The rooted plants of the littoral zone are in a sense benthos but are usually considered separately as *aquatic macrophytes*. Although there may be large numbers of microscopic invertebrates in the benthos, the more noticeable animals are such forms as sponges, flatworms, annelid worms, crustaceans such as crawfish (or crayfish), mollusk, and aquatic insect larvae. Bacteria are abundant. The food of the bacteria and most of the benthonic animals consists of the dead organic material, or *detritus*, that accumulates on the bottom.

The benthos of the profundal zone in productive lakes is particularly interesting because of the extreme low oxygen levels under which they may have to live during the summer stagnation period. Different animals show different mechanisms for dealing with this situation. Some larvae avoid it by migrating out of the profundal zone. The worms *Tubifex* and *Limnodrilus* and also the midge fly larva *Tendipes* have blood containing hemoglobin, the same oxygen-carrying pigment found in mammals which is especially efficient at transporting oxygen at low concentrations. The tendipedid larvae have the common name "bloodworms." Some bacteria of the profound zone are true anaerobes, requiring no oxygen in their respiration. Many animals of the profundal zone accumulate an *oxygen debt* during periods of little or no oxygen. Without oxygen to break glucose down to carbon dioxide, they break it down instead to lactic acid. The process is only about 5% as efficient as respiration in the presence of oxygen but it does provide some energy for the animal's needs. The lactic acid may be excreted, with a considerable loss of energy, or, later when oxygen is available again, it may be treated in such a way as to break it down to carbon dioxide (and water).

The *periphyton* are the microscopic or near-microscopic plants and animals that live attached or clinging to aquatic macrophytes, larger animals, rocks, or other surfaces projecting into the water. Included are diatoms and other types of algae, and such animals as hydras and the protozoans *Stentor* and *Vorticella*.

The *plankton* consists of the mostly microscopic plants (*phytoplankton*) and animals (*zooplankton*) that float in the water. Although most of the animals and some of the plants are capable of locomotion their movements are determined largely by water currents. The phytoplankton compose one of the two important producer groups in lakes (the other being the macrophytes); diatoms and blue-green and green algae are the important groups. The zooplankton organisms are mostly grazers on the phytoplankton, the most abundant being rotifers, copepods, ostracods, and water fleas.

Many of the limnetic zooplankton, especially the water fleas, show an interesting pattern of daily vertical migration. They migrate several meters each day, toward the surface at night and toward the bottom in daylight. The rate of ascent is usually a few centimeters per hour but may be a meter or more. The proximate factor in the vertical migration is light; the ultimate factor or factors are uncertain. It has been suggested that by coming up to feed at night the zooplankton (1) feed at the time when protein content of the algae is highest; (2) feed when predation is least; and (3) feed in waters of high temperature, allowing rapid feeding, while spending the rest of the day in waters of lower temperature, allowing more efficient growth.

The phytoplankton of lakes of temperate regions often show a characteristic pattern of abundance. After a winter minimum there is a spring peak of numbers corresponding with increasing day length and light intensity of spring. An increased nutrient supply in the epilimnion as a result of the circulation associated with the spring overturn and rising temperatures may also be involved. Diatoms are often important in the spring peak. A decline in late spring may result from a decrease in the nutrient supply through nutrients being tied up in dead bodies of plants and animals and deposited on the bottom. In the case of diatoms the limiting factor may be silica, of which their shells or *tests* are made. In nutrient-rich lakes a late summer "bloom" of blue-green algae often occurs but in nutrient-poor lakes algal populations may be low until fall. Another peak occurs about the time of the autumn overturn, when nutrients again are replenished in the upper layers of the lake. Individual species of algae may have yearly patterns of abundance very different from these overall patterns.

The *nekton* consists of the strongly swimming organisms, mostly fish but also including a few amphibians (such as the mud puppy) and birds (such as mergansers and loons). Depending on the species, fish may feed on detritus, plankton, small invertebrates, or other fish. One of the interesting features of fish communities, causing them to be more diverse ecologically than they would seem merely on the basis of species number, is the tendency for different sizes of the same species to use different foods. Small perch under 10 cm may eat mainly small zooplankton, larger perch may eat large zooplankton and some invertebrates from the bottom, and perch over 20 cm may eat mainly small fish. Changes in the specific habitat occupied may also occur, for example from littoral to limnetic.

The *neuston* is a group of remarkable organisms that are as they are because of another unusual property of water—it has the highest surface tension of any liquid except mercury. On land this allows the capillary rise of water in soils. In lakes, streams, and the ocean it produces a special habitat at the air-water interface, inhabited by the neuston. One set of neuston organisms lives on top of the surface film. Included are the long-legged water striders that actually walk on the water and the whirligig beetles that have two-parted eyes, the lower part suitable for the refractive index of water and the upper for the refractive index of air. Some other organisms use the underside of the water film, such as hydra, water fleas, and insect larvae such as mosquito wigglers.

The heterotrophs of lakes depend on three main sources of energy. These are production of organic matter in the lake by phytoplankton, production by macrophytes, and the import of organic matter from the surrounding basin. The imported material may vary from dissolved organic matter such as amino acids or carbohydrates to leaves, twigs, and tree trunks. One obvious type of food chain in the lake ecosystem begins with phytoplankton and ends with some top carnivore such as a trout. In between may be zooplankton, invertebrates, and small fish. But here, just as in many terrestrial ecosystems, such grazing food chains are relatively unimportant in the total energy metabolism of most lakes. Most organic material, whether originating as phytoplankton or macrophytes in the lake or as terrestrial plants carried into the lake, ends up on the bottom. Here it and the invertebrates and bacteria that work on it form the food chains through which most of the lake's energy input flows.

Succession and Eutrophication. The general fate of lakes is to be filled in by sediments from the surrounding uplands and also by the deposition of organic remains, and so to become dry land. The process may, however, be very slow. In much of the temperate part of the world, lakes tend to be unproductive, or *oligotrophic*, when they are first formed. Characteristically, they are deep and have low phytoplankton populations. With only a small amount of organic debris settling into a large volume of hypolimnion decomposition is not sufficient to exhaust the oxygen supply; thus, the benthos is diverse and there are cold-water fish, such as char and whitefish, living in the deep waters.

There is a tendency for lakes to become more productive, or *eutrophic*, as they age. Eutrophic lakes tend to have high algal populations, especially blue-greens, to show summer oxygen depletion in the hypolimnion and, consequently, to have few or no fish in the hypolimnion and a bottom fauna made up of only the forms tolerant of very low oxygen concentrations. It should be realized that the descriptions of oligotrophic and eutrophic lakes as given are like parts of a spectrum, and intermediate lakes do exist.

Natural *eutrophication* is related to an increased nutrient supply for the algae, especially of phosphorus and possibly also of nitrogen. Not all the causes of natural eutrophication are clear, however, nor is it clear that exactly the same processes are involved for all lakes. G. E. Hutchinson has suggested that after a short oligotrophic phase a steady state, or trophic equilibrium, tends to be established. At Linsley Pond, Connecticut, studied by Hutchinson and his colleagues, a steady state seemed to persist from seven or eight thousand years ago until the time of European settlement. The steady state of productivity may be partly based on a fairly constant nutrient input from the drainage basin.

Ponds and lakes receiving sewage and runoff from fields, fertilized lawns, or other high phosphate/nitrogen sources commonly show heavy "blooms" of blue-green algae in the summer which may look and smell unpleasant and which may interfere with swimming and boating. The other features of eutrophic lakes, such as oxygen depletion in the hypolimnion, may be accentuated. The large amount of organic debris produced tends to accumulate on the bottom and presumably the succession to dry land is speeded up. If Hutchinson's suggestion is correct, this extreme eutrophication connected with man's activities may have few natural parallels. In any case, it is clear that most lakes affected by increased nutrient input from man's activities can still be restored to approximately their previous productivity simply by stopping the input of sewage, fertilizer, or phosphate detergent.

The concentric zones of vegetation around the edges of lakes are usually regarded as representing successive seral stages in the last stages of succession from shallow water to land. The zones, from lakeward to landward, or from earlier to later, are often the following:

Submerged vegetation, with waterweeds, pondweeds, eelgrass, *Chara*, and bladderwort.

Floating vegetation, with water and pond lilies, pickerel weed, certain pondweeds, and duckweed.

Emergent vegetation, with cattail, reeds, rushes, bur reeds, arrowheads, and wild rice. This zone generally occupies the edge of standing water, from about 1.5 m depth onto the areas where the soil is merely damp.

Swamp shrub vegetation, with buttonbush, willows, and dogwoods. The zonation seems to be controlled primarily by water depth. Consequently, as the lake gets more shallow each zone tends to move outward and occupy the area previously held by the preceding one. For example, floating-leaved plants move out into areas earlier occupied by submerged vegetation while the area formerly dominated by floating-leaved plants is invaded by emergent vegetation.

STREAMS

Streams flow. That is not all that there is to know about them but it is by far the most important thing. The current sets the conditions under which stream organisms compared with other aquatic organisms must live. The water that was here in a stream yesterday is mostly far away today; whatever the organisms living in or along a stream do to the water is continually exported downstream.

Streams come in all sizes, from temporary trickles that flow only after rains to the miles-wide stretches of the great rivers such as the Mississippi and the Amazon. The classical term for flowing water is *lotic*. A basic subdivision is between *creeks* or *brooks*, which are less than 3 meters wide, and *rivers*, which are 3 meters or more. Streams of most sizes show two basic habitats, *rapids* or *riffles* and *pools*.

It is the riffle that is the true stream habitat. Here the water is swift-flowing over rocks or pebbles. Because of the turbulent flow air is continually mixed with the water, so the oxygen level is always high. The plants of the riffles are mostly algae that grow attached to the rocks. They may be filamentous forms that grow as cushion- or crust-like mats, or they may be sessile diatoms that live as periphyton on other algae or attached to rocks and pebbles. The vegetation in streams that most people would call "moss" is usually algae but there are a few aquatic mosses that are important in some streams.

Animals of the riffles keep from being swept away by the current through three main mechanisms: they attach themselves to solid surfaces, swim strongly, or avoid the current. Many animals show some combination of these methods.

Sponges and bryozoans behave essentially like plants—they are sessile, living attached to rocks or to dead trees in the water. However, many other invertebrate animals move around but cling to rocks or vegetation using hooks, claws, or suckers. The limpet is a characteristic swift-water animal, a peculiar snail which clings to rock surfaces by means of its muscular foot, and also diminishes the force of the current by its low conical shape that resembles a chinese coolie's hat. Water pennies, the larvae of a certain kind of beetle, are almost flat; they cling to the surface of rocks where the current is slow. Blackfly larvae spin silk threads to which they attach themselves by a circle of hooks on their posterior end. Mayfly naiads and many other immature insects occurring in the riffle are streamlined to reduce drag, and have legs adapted for clinging to the substrate. Many caddisflies live in heavy cases constructed of sand, pebbles, or twigs cemented together.

In addition to having shapes that reduce drag, animals may avoid the current by actually going where there is little or no current. They may go under rocks or even down into the rubble on the bottom. Also, within dense mats of vegetation, the current may be practically zero.

Several kinds of invertebrates such as crawfish and amphipods can swim, but fish are the stream organisms that can swim strongly enough to maintain their position readily. Combined with the ability to swim, fish show an orientation response to current, or *rheotaxis*. Fish and many stream invertebrates will face into and move against a current. Some fish seem to use landmarks along the shore as cues for maintaining their position in the stream but darters, perhaps the most characteristic riffle fishes, seem to use tactile stimuli from the bottom.

The current in certain parts of the stream, including shallow margins, backwaters, and pools, is much reduced. Pools ordinarily have water entering at the upstream high side and spilling out at the downstream low side, with no real through-flow. With the lowered velocity, sand, silt, and clay carried by the stream is deposited, so that these areas tend to have sand or mud bottoms. Many of the organisms living there, especially in the mud-bottomed pools, are the same as those occurring in ponds. This includes some of the same marsh plants such as cattail, arrowleaf, and smartweed. Watercress is a characteristic stream emergent rarely, if ever, found in ponds.

Some characteristic pool animals include burrowing mayfly naiads, tendipedid larvae, tubificid or limnodrilid worms, and various clams, in the benthos. The nekton consists mostly of fish, some the same species as occur in ponds but with several distinctive stream fish. A neuston of water striders and whirligig beetles is often well developed.

Plankton in general is poor in streams, except in the larger rivers. Here sizable populations of plankton may develop in a mass of water as it moves toward a large lake or the ocean. Most characteristic are certain diatoms, rotifers, water fleas, and copepods. Plankton is rare in smaller streams; a plankton net put into the water will sieve out only a few true planktonic organisms. It will, however, pick up bottom dwellers such as mayfly naiads and the larvae of caddisflies, blackflies, and midges.

It is now clear that such downstream **drift** is so prevalent that it must be a phenomenon of fundamental importance in streams. Although much is still uncertain, it is known that drift is more pronounced at night than in the day, and in spring and summer rather than at other seasons. For some organisms, such as mayflies, drift seems to be part of cyclical, migratory phenomena in which the immature stages drift downstream and the adults, upon emergence, fly upstream and lay their eggs in the headwaters, allowing the cycle to continue. The evolutionary advantage of drift apparently is that it may bring the immature insects to less crowded situations. One effect of this in the functioning of the ecosystem is that areas of a stream depopulated by floods are quickly recolonized. The same is true of sections of streams made barren by heavy pollution.

Most streams are strongly heterotrophic systems. The high populations of animals that they sustain are parts of food chains based mainly on imported organic matter rather than upon the small amount of plant matter produced in the stream. The imported organic matter enters the stream usually as leaves but also as twigs, fruits, and other plant parts. The large pieces of organic matter are quickly colonized or attacked by bacteria, fungi, and animals that K. W. Cummins has called "shredders." Included are various caddisflies, crane flies, and stone flies. Other organisms use the smaller detritus fragments that result from the activities of these organisms and organic compounds leached out of the leaves. Many of the detritus feeders may obtain their nutrition as much from the bacteria they ingest as from the leaves or leaf fragments. There are, of course, predators that feed on the detritus and plant feeders; among these are dobsonflies, some stone flies, and certain fish such as sculpins.

In small shaded forest streams as much as 99% of the organic material may be imported (*allochthonous*), and 1% or less produced by in-stream photosynthesis (*autochthonous*). In large streams the dependence on allochthonous input is less, but too few streams have been studied to provide even rough quantitative estimates.

THE OCEANS

The oceans are vast, and so is *oceanography*, the science of the oceans. Oceans occupy just over 70% of the earth's surface and have an average depth of about 4000 meters, or something more than two miles. The deepest areas are trenches in the western Pacific more than 10,000 meters deep.

One of the important ways that the oceans differ from fresh-water habitats is in *salinity*, or saltiness. Ocean water contains about 3.5% salt by weight, generally expressed as 35 parts per thousand (rather than 3.5 parts per hundred) and written 35%oo. The salinity of fresh water, by contrast, is mostly below 2%oo. To an inlander the saltiness of the ocean might seem to make it an extreme environment, to be populated by a biota of low diversity but in fact the opposite is actually the case. Life originated in the oceans and evolution went on there more than two billion years before plants and animals began to evolve the traits allowing them to use such inhospitable habitats as dry land and fresh water. Only

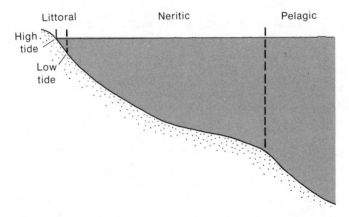

Figure 5–24. Major zones of the ocean.

a few groups such as the seed plants, the insects, and the birds and mammals are better represented outside the ocean than in it. There are whole phyla such as the comb-jellies, brachiopods, chaetognaths, and echinoderms that do not occur in fresh water at all.

Very broadly, the ocean consists of three zones (Fig. 5–24): a *littoral zone* that is approximately the area between the lines of high and low tide, a *neritic zone* that is the portion of the ocean over the continental shelf with depths down to about 200 meters and, by far the largest, the *pelagic zone* of the open ocean.

Littoral Zone. The littoral zone of the ocean is a habitat of physical extremes, changing from aquatic to terrestrial as the tide goes out, twice a day. Accompanying this are changes in temperature and several other physical factors. Moreover, the organisms there are subjected to the pounding of incoming waves and the scouring of the water as the waves recoil.

It is usual to recognize three zones on both sandy and rocky shores. These are the *intertidal* zone itself, a *supratidal* zone above it which is covered only by the highest tides or influenced by spray, and the *subtidal* zone below the line of average low tide. The organisms inhabiting the three zones are almost entirely different on rocky and sandy shores.

Many of the organisms of rocky shores are sessile, permanently attached to the rock. The supratidal zone is a spray region where black-colored blue-green algae grow on the rocks and are fed upon by small snails called periwinkles. Below this the intertidal zone, despite the harsh physical conditions, has a large and varied fauna. Barnacles, snails, limpets, oysters, mussels, sea anemones, chitons, and sea urchins are some of the animals that occur attached to the rocks, in crevices, or in the masses of brown and red algae. Below the average low tide line, in the subtidal zone, grow large brown algae called kelps. Here too are a great variety of animals, including many that move up into the intertidal zone when it is water-covered. Typical animals of the subtidal zone include sea anemones, starfish, sea urchins, polychaete worms, sea cucumbers, nudibranchs, crabs, and tunicates.

On sandy (and muddy) shores the substrate is unstable; here burrowing rather than attachment is the predominant adaptation. Animals of the supratidal zone include ghost crabs and amphipods which shift their activities into the intertidal zone when the tide is out. In the intertidal zone are large populations of clams, polychaete worms, isopods, and amphipods, crabs, and other crustaceans. Most of these animals are hidden in burrows when the tide is out but are active when submerged. In the subtidal zone occur fish such as flounders and killifish, some of which move up into the intertidal zone at high tide, and sand dollars, clams, snails, and blue crabs.

On sandy beaches, unlike rocky beaches, there is little production of organic matter by plants. Consequently grazing and plankton feeding are not very important; most of the organisms are detritus feeders.

Estuaries are coastal bodies of water connected with the ocean, such as tidal marshes, tidal rivers, and the bodies of water behind barrier islands. Typically, estuaries are highly productive at both the producer and consumer levels, for a variety of reasons. One is the action of the tides, bringing nutrients and carrying away waste products. Another reason for the high productivity is the wide variety of producers that can live in the estuarine habitats, from plankton to seaweeds to the very important emergent marsh grasses. Some of the organic matter produced in estuaries is exported into the adjacent ocean.

Neritic and Pelagic Zones. The important producers are phytoplankton, except in the shallowest parts of the neritic zone. Diatoms, especially in cooler waters, and dinoflagellates are numerous; even more important may be *microflagellates,* too small to be collected by plankton nets but extremely abundant, especially in the open ocean.

Phytoplankton is fed upon by zooplankton, and smaller zooplankton may be eaten by larger. The zooplankton is diverse, especially in neritic regions where it is enriched by the motile larvae of sedentary animals of the bottom and sessile animals of the littoral zone. Examples of the animals which live only temporarily in the plankton, as larvae, are barnacles, sea urchins, polychaete worms, and crabs. The permanent plankton is diverse in itself and includes more larger forms than are ordinarily important in fresh water. Some of the important groups are copepods, the good-sized crustaceans called "krill," jellyfish and comb-jellies, and protozoans such as foraminiferans and radiolarians.

Nekton, mostly feeding on the zooplankton, consists of such fish as herring, menhaden, sardines, squid, and baleen (or whalebone) whales. Higher-order carnivores among the nekton include toothed whales, sharks, and birds. Pelagic birds are mostly restricted to albatrosses, shearwaters, petrels, and penguins; in neritic regions are added gulls, terns, auks, pelicans, gannets, and many others.

The depth to which photosynthesis exceeds plant respiration varies, but it is probably rarely more than about 30 meters in the more cloudy waters of the neritic zone nor more than about 200 meters in the clearest waters of the pelagic zone. Organisms at greater depths are dependent on import of organic matter originally produced above. In shallow waters much of this import may occur by the sinking of dead bodies of plankton and other organisms. In deeper waters the organic remains of organisms from near the surface probably decay before they reach the bottom. The way in which the organic matter that supports the benthos and the deep water nekton reaches the bottom is not wholly understood. The aggregation of dissolved organic matter into particles, activities at all levels of plankton able to use dead organic material, and transport along the bottom from shallower regions may all be important.

Populations of the benthos decline with increasing depth. Some important forms are clams, other molluscs, sea urchins, and polychaete worms in shallower water and brittle stars, sea cucumbers, sea lilies (or crinoids), brachiopods, and glass sponges in deep water. Smaller organisms are also present. Most benthonic animals are detritus feeders but some, such as the brittle stars, are carnivores. Many fishes of the ocean depths are also carnivores. Here is where such bizarre forms as the gulpers, swallowers, and angler fish occur.

It is worth noting that currents occur throughout the ocean and, at least partly as a result, oxygen depletion such as at the bottom of lakes is rare in the sea. The fauna is not one adapted to low oxygen levels. Some of the more prominent adaptations are to the soft nature of the bottom, such as long arms, fins, or spines for support, and to the constant darkness. Many of the fish and squids of the depths have luminescent organs where light is produced in the same way as in a lightning bug's abdomen. The light is used in various ways. Some organisms seem to have searchlights, others have luminous baits to lure prey, and in some cases the luminous structures may serve for sex or species recognition. There are few obvious special adaptations to pressure, despite the fact that the pressure at an average depth of the ocean bottom is more than 350 atmospheres. One feature shared by the organisms of the depths is the absence of gas-filled cavities such as swim bladders in fish. Beyond this, it may be sufficient that pressure within and outside the organisms are the same.

Productivity. Popular literature that talks about the "infinite resources of the sea" has now nearly disappeared. It has become clear that fish and even algae as food are no more infinite than the blue whale which has been hunted virtually to extinction. Even so it still surprises most people when they discover just how unproductive most of the ocean is. A reasonable figure for net primary production in the open ocean is probably 500 to 600 kcal per square meter per year, similar to semidesert communities on land.

The factor limiting production in the open ocean appears to be nutrient supply, generally phosphorus. The problem is the continual loss of nutrients from the photosynthetic zone. This occurs mainly through organisms dying and sinking, eventually to the bottom. Furthermore, the bottom is far from the top, and there is nothing which corresponds exactly to the overturns of fresh-water lakes which replenish the nutrient supply at the top. If all nutrient regeneration were postponed until dead organisms reached the bottom of the ocean, the nutrient supply would probably be much poorer than it actually is. In fact, considerable decay goes on as these dead bodies slowly sink and, furthermore, there is direct use by phytoplankton of zooplankton excretions, such as ammonia.

In a few situations in the ocean nutrient levels are kept high, and these areas are highly productive. Of particular importance to man are areas of *upwelling*, coastal areas where nutrient-rich waters from lower levels circulate upward by the action of currents carrying surface waters away from shore. Probably the best known of such areas is that of the Peru current. Here, in an area of the Pacific Ocean about the same size as Lake Michigan, man harvests some 10.5 million metric tons of anchovies per year. The net annual production by the phytoplankton, especially diatoms, at the base of the food chains here is clearly very high, possibly on the order of 4000 or 5000 kcal per square meter per year. Estuaries and coral reefs may have even higher levels of productivity.

PALEOECOLOGY AND HISTORY OF THE BIOMES

Unlike the molecular biologist, the chemist, or the physicist, knowledge of present conditions is often not enough for the ecologist. The past is important. This may be the recent past—a game biologist may need to know what happened to a population last winter to understand what is likely to happen to it this summer. Or it may go back a little farther—a plant ecologist studying the prairie may reach some wrong conclusions unless he knows whether the past few years have been wet or dry. Understanding why communities are organized or distributed as they are may require a knowledge of events which occurred thousands or millions of years ago. The field of study dealing with the interrelations of organisms and environment in geological time is called *paleoecology*.

Like any other kind of ecology, paleoecology can study individuals, populations, or communities. There is, for example, an interesting population study of the extinct cave bear in Scandinavia. We will focus on community ecology, specifically the history of the North American biomes since about 65 to 70 million years ago.

There are three main episodes in this history. To begin with, climates were milder than today and tropical forests grew in much of what is now the U. S. Climates then began to get cooler and dryer and the vegetation changed, so that much of the U.S. was occupied by temperate forest and grassland. Finally the Ice Age arrived, with glaciers advancing and receding, and the vegetation changed greatly with each advance and recession.

At the beginning of the Paleocene (Table 5–6) the climate in North America was probably warmer and there was probably less difference between summer and winter. Ralph W. Chaney and Daniel I. Axelrod have described three great geofloras which occurred in North America at this time, the Neotropical-Tertiary, the Arcto-Tertiary, and the Madro-Tertiary Geofloras. The southern part of the continent, extending north to what are now the states of Washington and Colorado, was occupied by the Neotropical-Tertiary Geoflora (Fig. 5–25). The fossil leaves and fruits found in rocks laid down at this time show that the plants present were similar to those now found in tropical and subtropical forest and savanna.

The northern part of the continent was occupied by communities of the Arcto-Tertiary Geoflora. At this time there was still a land connection between North America and western Europe across the North Atlantic, and this geoflora ranged all across North America and Eurasia at higher latitudes. In its southern portions this was temperate deciduous forest, evidently similar to the mixed mesophytic forests of the present day. Many of the same kinds of trees were present, probably not the same species but close relatives. There were maples, beech, chestnut, ash, wal-

Table 5–6 Geological Timetable

Era	Period	Epoch	Years Ago
Cenozoic	Quaternary	Recent	5,000
		Pleistocene	2,000,000
	Tertiary	Pliocene	7,000,000
		Miocene	26,000,000
		Oligocene	38,000,000
		Eocene	54,000,000
		Paleocene	65,000,000
Mesozoic	Cretaceous		136,000,000
	Jurassic		190,000,000
	Triassic		225,000,000
Paleozoic	Permian		280,000,000
	Carboniferous (Pennsylvanian and Mississippian)		345,000,000
	Devonian		395,000,000
	Silurian		430,000,000
	Ordovician		500,000,000
	Cambrian		570,000,000
Precambrian			5,000,000,000

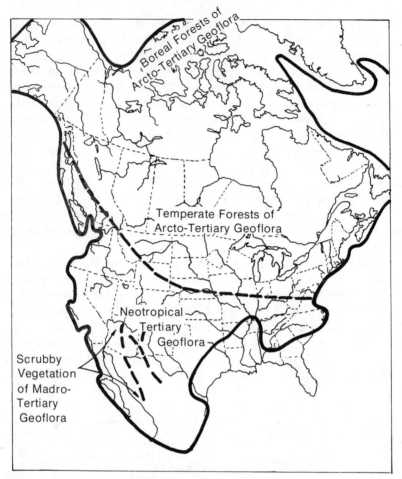

Figure 5–25. Diagrammatic representation of the vegetation of North America in the Paleocene, approximately 65 million years ago. At this time there was still a North Atlantic connection with Eurasia through Greenland but the Bering Straits connection had not yet been established. The central American isthmus connecting North and South America was not yet in existence but the southeast coast of North America and the northeast coast of South America were fairly close together. The coastal plain of the southeast U.S., north to about Cairo, Illinois, was still under water and, not long before, the area of the Great Plains had been submerged under a broad, shallow sea.

nut, oaks, basswood, elms, and many others. To the north, perhaps in what is now northern Canada, conifers such as pine and spruce increased but there were large numbers of deciduous trees even in that region.

One of the most characteristic members of the Arcto-Tertiary forests was the dawn redwood, or *Metasequoia*. This plant, related to the present-day sequoia and redwood of the west coast, was known only as a fossil for many years and then was found in China in 1946. It was growing in valleys with chestnut, oak, sweet gum, cherry, beech, and spice bush or, in other words, in an association much the same as found in fossil beds 65 million years old.

The third geoflora was a newly developing one, the Madro-Tertiary Geoflora, in the drier, cooler highlands of the southwest U.S. and Mexico. The plants included such types as live oaks, pines, junipers, yucca, and mesquite.

This was the situation in the Paleocene and Eocene. By the end of the Oligocene, 26 million years ago, a climatic deterioration had begun, with climates becoming cooler and drier. This produced two general trends. First, there was a southward shifting of communities, so that by the end of the Miocene the Neotropical-Tertiary Geoflora had retreated south of the U.S. and the Arcto-Tertiary Geoflora had spread southward and occupied much of North America. We need to guard against an anthropomorphic view of these geofloras packing their bags and heading south. The actual movements occurred because conditions at the north edge of the range of the tropical species became unfavorable so that these species could no longer survive or reproduce. As these died out more northerly species found conditions favorable and were able to grow, thereby extending their range southward.

The second effect resulting from the cooling and drying climatic trend was the spreading out of the Madro-Tertiary Geoflora from its place of origin, and the development of various vegetation types such as chaparral, woodland, and desert, adapted to dry to very dry conditions.

Sometime in the Miocene temperate deciduous forest derived from the Arcto-Tertiary Geoflora was extremely widespread, occupying large areas of North America and Eurasia. This is the basis for the similarities between the forests of western Europe, eastern Asia, and eastern North America; they were once part of a continuous deciduous forest vegetation type. Subsequent events in the last 20 million years have led to the fragmentation of the deciduous forest, complete extinction of some lines, and extinction of some lines in several regions but not in others.

Continued cooling and drying led to the development of forests dominated by evergreens in the northern part of the continent and the further spread of arid-land communities, including grassland, into the interior of North America. The trend of greater dryness in western North America was produced by the uplift of mountains in the West; these mountains produced a rain shadow in the interior. Areas of the Great Plains changed from forest to savanna to grassland over a great many millions of years and the animals present also changed, by emigration, immigration, and evolution.

The evolution of horses is a favorite example for the evolution section of beginning biology textbooks. Here is the place and the time in which that evolution occurred. In the Eocene, horses were four-toed little animals about the size of fox terriers who browsed on tree leaves. There were several lines of evolution, one of which, by the early Pleistocene, led to the big creatures we ride or bet on, having a single toe with a hoof on it and teeth suitable for cropping and chewing up grass (Fig. 5–26).

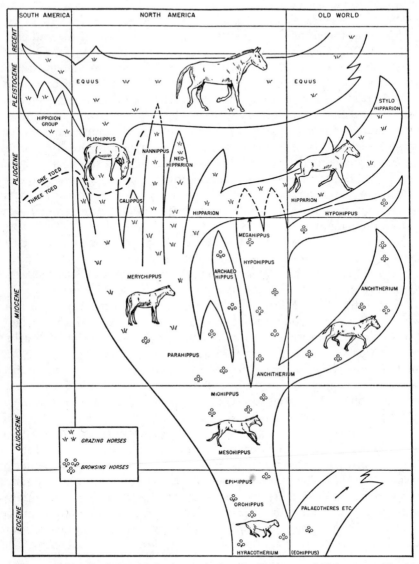

Figure 5–26. Evolution of the horse during the Tertiary and Pleistocene. The main lines are shown. The horse evolved in North America but migrated to Eurasia. Like most other large mammals of North America it became extinct in the Pleistocene. (From G.G. Simpson, *Horses,* New York, Oxford University Press, 1951, p. 115.)

By the beginning of the Pleistocene, the distribution of the biomes may have been rather similar to what they are now in North America. There were probably differences, but at first glance a biome map in a late Pliocene ecology text might have looked much like the one in this book (Fig. 5–27). However, there were differences in the composition of the communities as compared with today. One difference is that the Pleistocene communities had more large animals. The fossils found in the tar pits at Rancho LaBrea in Los Angeles County, California, are examples. The tar pits are places where asphalt comes to the surface and animals become trapped, like Brer Rabbit, probably when they come to drink at water holes or to prey on some other animal there. The bones preserved in the asphalt have been studied by Hildegard Howard, Chester Stock, and others. Based on their work, some of the large animals that lived in the vegetation types around the Los Angeles valley in the late Pleistocene included herbivores such as 4-foot–tall ground sloths, horses, camels, mammoths, mastodons, antelopes, and peccaries, and predators and scavengers such as the saber-toothed cat, dire wolf, short-faced bear (larger

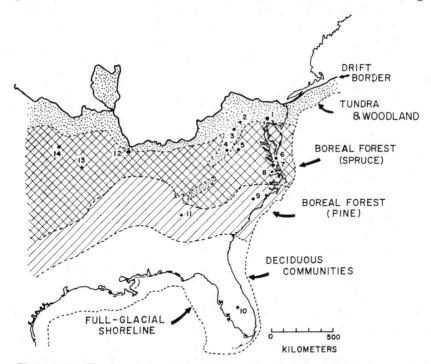

Figure 5–27. The distribution of vegetation and associated animals in eastern North America during "full-glacial" time, perhaps 20,000 years ago, may have been something like this; however, a great deal is still unclear about the location and composition of communities. Large areas of land now covered by the ocean were exposed at this time because of low ocean levels resulting from the large amount of water tied up in glacial ice. (From D. R. Whitehead, "Late-Wisconsin vegetational changes in unglaciated North America," *Quaternary Research*, 3(4):628, 1973.)

than the grizzly bear), and vultures with a wing span of 14 feet. Large animals of these species or others also occurred elsewhere in North America throughout most of the Pleistocene but then disappeared, rather abruptly, 10,000 to 11,000 years ago.

The climatic deterioration that began in the Oligocene reached a climax with the continental glaciation of the Pleistocene. Four or more times glaciers came down from the north as ice sheets hundreds to thousands of feet thick. Communities occupying the land over which they advanced were obliterated. Plants and animals were shifted south as the result of the cooler climates. We know, as yet, relatively little about the early glaciations. About the last period of glaciation, the Wisconsin, we know much more.

Wisconsin glaciation began about 70,000 years ago and was at its peak about 20,000 years ago. After this, the glaciers melted back quickly. The front of the ice was sitting in southern Michigan about where I am now possibly 18,000 years ago. By 9,000 or 10,000 years ago the glacier was north of the Great Lakes basin and by 4,000 or 5,000 years ago it was gone from the North American continent. Most geologists believe that we are now in an interglacial period, with the next glaciation due somewhere between a few hundred to many thousand years hence.

It is clear that there was considerable southward movement of both plants and animals during glaciation (Fig. 5–27). From the fossil pollen in lake sediments and bogs we know that jack pine and spruce grew in central Georgia. Bones of musk ox have been found in Indiana and lemmings in southern Pennsylvania. It is not clear whether vegetation occurred in zones such as those today but simply shifted south, or whether the landscape was more complex. If the second hypothesis is true, some temperate deciduous forest may have survived close to the glacial border in favorable microclimates, such as sheltered south-facing slopes; however, the question is far from resolved.

With glacial recession, large areas of bare ground were left. Of course, pioneer vegetation developed first. Succession then tended to proceed to tundra. Past this point, the marked vegetation changes that occurred were the result of the changing climate, with some modification by the immigration rate of various species back into the glaciated areas. A spruce woodland followed the tundra, and this developed into a spruce forest. Around 11,000 years ago over a rather wide area spruce forest gave way to pine forest. Pines gave way to hardwoods, 8,000 or 9,000 years ago (at the level of southern Michigan) and since that time the prevailing communities have been hardwood forests of one type or another. There have been one or more periods warmer than now during that time span, and there seems to have been a warm, dry period around 3,000 years ago when grassland communities and oak-hickory forest expanded at the expense of mesic forest.

The fossils of mastodons, mammoths, giant beaver, woodland musk ox and such disappeared around 10,000 to 11,000 years ago. By 4,000 years ago, and probably well before, the fauna was a modern one of deer, raccoon, teal, coot, and mallard. What happened to the large Pleistocene animals? P. S. Martin has pointed to some seemingly peculiar features of their extinction: unlike the usual course of extinction, it did not involve replacement by other kinds of organisms having similar niches, and it was mainly the large animals that were involved. He has assembled data suggesting that large-scale extinctions occurred on all the continents at about the time primitive man appeared there. His conclusion is that

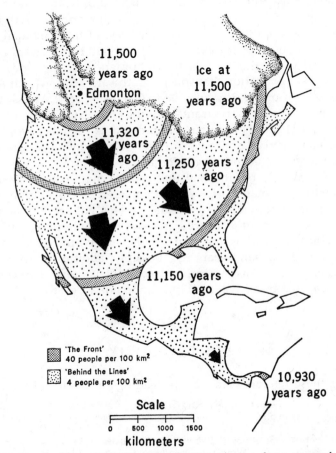

Figure 5–28. One hypothesis for the extinction of many large mammals in North America 10,000 to 11,000 years ago is overkill by Paleoindians. This diagram suggests how it might have happened. A small band arrived in southern Canada about 11,500 years ago. They and their descendants set off a "front" of human occupation that spread southward. They were big-game hunters, killing off mammoths and mastodons in their neighborhood and then moving on to better hunting grounds. To the rear of the front, smaller populations of humans may have remained, living on smaller game. Just how long it would have taken to exterminate the big game all the way to the Gulf of Mexico depends on such factors as population growth rates for mammoths and man and how strongly mammoth and mastodon were preferred as food. (From P.S. Martin, "The Discovery of America," *Science*, 179:972, 1973. Copyright 1973 by the American Association for the Advancement of Science.)

primitive man, in our case Paleoindians, caused the extinctions, primarily through hunting the animals for food. Large predators might not have been hunted but might have become extinct when the large herbivores they fed on were exterminated.

Some have had difficulty accepting the picture of a small number of primitive humans exterminating millions of mastodons, mammoths, and other large animals. Extinction, however, is the outcome predicted by computer models based on perfectly reasonable assumptions about population sizes of hunter and hunted, migration and kill rates, and population growth rates. Humans presumably arrived in North America across the Bering land bridge. The models begin with a band of 100 men and women who followed the ice-free corridor east of the Canadian Rockies from Alaska to southern Alberta, arriving about 11,500 years ago (Fig. 5–28). According to the models these hunters and their descendants formed a front of extinction that spread southward across North America, reaching the Gulf of Mexico only 300 to 500 years later. At that time the human population along the front, running from New England to Louisiana and into Mexico, might have numbered as high as a few hundred thousand. In its spread southward it would have killed on the order of 100 million mammoths and other big game.

We have already pointed out that countless animals and plants became extinct before man ever appeared on earth. The overkill hypothesis has been criticized on several grounds—it does not seem to account for the large-scale extinctions of birds at about the same time, and there is some evidence suggesting that man may have been in North America as early as 20 or 30 thousand years ago. Possibly, additional factors such as climatic change or disease were involved in Pleistocene extinctions. But it seems clear that primitive man, like modern man, was a potent ecological force.

BIBLIOGRAPHY

Following these references is a list of readings for various biomes.

Axelrod, Daniel I. "Evolution of the Madro-Tertiary geoflora," *Bot. Rev.*, 24:433–509, 1958.
Brewer, R. "Stability in bird populations," *Occas. Papers Adams Ctr. Ecol. Studies*, No. 7:1–12, 1963.
———. "Bird populations of bogs," *Wilson Bull.*, 79:371–396, 1967
———, and Merritt, P. G. "Wind throw and tree replacement in a climax beech-maple forest," *Oikos*, in press.
Brookhaven National Laboratory. "Diversity and stability in ecological systems," *Brookhaven Symp. Biol.*, No. 22:1–264, 1969.
Chaney, Ralph W. "Tertiary centers and migration routes," *Ecol. Monogr.*, 17:139–148, 1947.
———. "The bearing of the living *Metasequoia* on problems of Tertiary paleobotany," *Proc. Natl. Acad. Sci.*, 34:503–515, 1948.
Clements, F. E. *Plant Succession and Indicators.* New York, H. W. Wilson, 1928.
———. "Nature and structure of the climax," *J. Ecol.*, 24:252–284, 1936.

Cowles, Henry Chandler. "The ecological relations of the vegetation on the sand dunes of Lake Michigan," *Bot. Gaz.*, 27:95–117, 167–202, 281–308, 361–391, 1899.

Curtis, John T. *The Vegetation of Wisconsin.* Madison, University of Wisconsin Press, 1959.

Davis, M. B. "Pleistocene biogeography of temperate deciduous forests," *Geoscience and Man*, 13:13–26, 1976.

Denevan, W. M. "Development and the imminent demise of the Amazon rain forest," *Prof. Geogr.*, 25:130–135, 1973.

Dort, W., Jr., and Jones, J. K., Jr. *Pleistocene and Recent Environments of the Central Great Plains.* Lawrence, Kansas, University of Kansas Press, 1970.

Drury, William H., and Nisbet, I. C. T. "Succession," *J. Arnold Arboretum*, 54:331–368, 1973.

Elton, Charles S. *The Ecology of Invasion by Animals and Plants.* London, Methuen, 1958.

Gleason, H. A. "The individualistic concept of the plant association," *Am. Midl. Nat.*, 21:92–110, 1939.

Goodall, D. W. "The continuum and the individualistic association," *Vegetatio*, 11:297–316, 1963.

Henderson, L. J. *The Fitness of the Environment.* New York, Macmillan, 1913.

Hutchinson, G. E. "Nitrogen in the biogeochemistry of the atmosphere," *Am. Sci.*, 32:178–195, 1944.

Kapp, R. O. "Late Pleistocene and postglacial plant communities of the Great Lakes region," in R. C. Romans, ed., *Geobotany.* New York, Plenum Press, 1977, pp. 1–27.

Kenoyer, L. A. "Forest distribution in southwestern Michigan as interpreted from the original land survey (1826–32)," *Papers Mich. Acad.*, 11:211–217, 1934.

Loucks, O. L. "Evolution of diversity, efficiency, and community stability," *Am. Zool.*, 10:17–25, 1970.

Martin, P. S. "The discovery of America," *Science*, 179:969–974, 1973.

———, and Wright, H. E., Jr., eds. *Pleistocene Extinctions.* New Haven, Yale University Press, 1967.

May, R. M. *Stability and Complexity in Model Ecosystems,* 2nd ed. Princeton, Princeton University Press, 1974.

McIntosh, Robert P. "The continuum concept of vegetation," *Bot. Rev.*, 33:130–187, 1967.

———. "H. A. Gleason—'individualistic ecologist' 1882–1975: his contributions to ecological theory," *Bull. Torrey Bot. Club*, 102:253–273, 1975.

McNaughton, S. J. "Diversity and stability of ecological communities: a comment on the role of empiricism in ecology," *Amer. Nat.*, 111:515–525. 1977.

Merriam, C. Hart. "Results of a biological survey of the San Francisco Mountain region and the desert of the Little Colorado, Arizona," *North Am. Fauna*, No. 3, 1–136, 1890.

Odum, Eugene P. "The strategy of ecosystem development," *Science*, 164:262–270, 1969.

Olson, J. S. "Rates of succession and soil changes on southern Lake Michigan sand dunes," *Bot. Gaz.*, 119:125–170, 1958.

Raynal, D. J., and Bazzaz, F. A. "Interference of winter annuals with *Ambrosia artemisiifolia* in early successional fields," *Ecology*, 56:35–49, 1977.

Regal, P. J. "Ecology and evolution of flowering plant dominance, *Science*, 196:622–629, 1977.

Shelford, Victor E. *Animal Communities in Temperate America.* Chicago, University of Chicago Press, 1913.

———, ed. *Naturalist's Guide to the Americas.* Baltimore, Williams and Wilkins, 1926.

———. *The Ecology of North America.* Urbana, Ill., University of Illinois Press, 1963.

Simpson, George Gaylord. *Horses.* New York, Oxford University Press, 1951.

Steinbeck, J. *The Grapes of Wrath.* New York, Viking Press, 1939.

Stock, Chester "Rancho La Brea, a record of Pleistocene life in California," *Los Angeles County Mus. Sci.*, Ser. No. 20, 1–81, 1961.

Vestal, A. G. "Why the Illinois settlers chose forest lands," *Trans. Ill. Acad. Sci.*, 32:85–87, 1939.

Weaver, John E. "Return of midwestern grassland to its former composition and stabilization," *Occas. Papers Adams Ctr. Ecol. Studies*, No. 3:1–15, 1961.

Whittaker, R. H. "A consideration of climax theory: the climax as a population and pattern," *Ecol. Monogr.*, 23:41–78, 1953.

Yarnell, R. A. "Aboriginal relationships between culture and plant life in the Upper Great Lakes Region," Ann Arbor, University of Michigan Museum of Anthropology, Anthropological Paper No. 23, 1–218, 1964.

BIOMES AND AQUATIC ECOSYSTEMS: SELECTED READINGS

Most of these sources are technical books and articles but I have also tried to list some popular ones. These often give a better idea of the look, feel, and smell of an area than can be gotten from a species list or a table of temperatures. For all the biomes, the appropriate chapters in S. Charles Kendeigh's *Ecology with Special Reference to Animals and Man* (Englewood Cliffs, Prentice-Hall, 1974) are excellent sources of information, and the same is true of V. E. Shelford's *The Ecology of North America* (op. cit.). My debt to the thinking and writing of both men will be evident. Popular accounts of biomes with color photographs are included in the McGraw-Hill series *Our Living World of Nature* and the Time-Life series the *Life Nature Library*.

Temperate Deciduous Forest

Braun, E. Lucy. *Deciduous Forests of Eastern North America*. Philadelphia, Blakiston, 1950.
Curtis, John T. *The Vegetation of Wisconsin*. Madison, University of Wisconsin Press, 1959, Chapters 5–12.
Rand, A. L., and Rand, R. M. *A Midwestern Almanac*. New York, Ronald Press, 1961.
Rawlings, Marjorie Kinnan. *Cross Creek*. New York, Scribners, 1942.
Reichle, David, ed. *Studies in Ecology I: Analysis of Temperate Forest Ecosystems*. Berlin, Springer-Verlag, 1970.
Thoreau, Henry David. *Walden*. Boston, Ticknor and Fields, 1854.
Twomey, Arthur C. "The bird population of an elm-maple forest with special reference to aspection, territorialism, and coactions," *Ecol. Monogr.*, 15:173–205, 1945.
Williams, A. B. "The composition and dynamics of a beech-maple climax community," *Ecol. Monogr.*, 6:318–408, 1936.

Temperate Grassland

Brewer, Richard. "Death by the plow," *Natural History*, 79:28–35, 110, 1971.
Carpenter, J. R. "The grassland biome," *Ecol. Monogr.*, 10:617–684, 1940.
Craig, W. "North Dakota life: plant, animal, and human," *Bull. Am. Geo. Soc.*, 40:321–332, 401–415, 1908.
Dort, W., Jr., and Jones, J. K., Jr. eds. *Pleistocene and Recent Environments of the Central Great Plains*. Lawrence, Kansas, University of Kansas Press, 1970.
Peattie, D. C. *A Prairie Grove*. New York, Simon and Schuster, 1938.
Weaver, J. E. *North American Prairie*. Lincoln, Nebraska, Johnsen, 1954.

Tundra

Billings, W. D. "Arctic and alpine vegetations: similarities, differences, and susceptibility to disturbance," *BioScience*, 23:697–704, 1973.
Bliss, L. C., et al. "Arctic tundra ecosystems," *Ann. Rev. Ecol. Syst.* 4:359–399, 1973.
Ives, J. D., and Barry, R. G., eds. *Arctic and Alpine Environments*. London, Methuen, 1974.
Murie, Adolph. *The Wolves of Mount McKinley*. Washington, D.C., U.S. Dept. Interior, Nat. Park Service, Fauna Ser. 5, 1944.
Pitelka, Frank A., Tomich, P. Quentin, and Treichel, F. W. "Ecological relations of jaegers and owls as lemming predators near Barrow, Alaska," *Ecol. Monogr.*, 25:85–117, 1955.
Scherman, K. *Spring on an Arctic Island*. New York, Little, Brown, 1956.

Coniferous Forest

Harvey, L. H. "A study of physiographic ecology of Mt. Katahdin, Maine," *Univ. Maine Studies*, 5:1–50, 1903.

Hoover, Helen. *The Long-Shadowed Forest*. New York, Thomas Y. Crowell, 1963.

Korling, Torkel. *Wild Plants in Flower: The Boreal Forest and Borders*. (Notes on species by E. G. Voss.) Dundee, Illinois, privately published, 1973.

Shelford, V. E., and Olson, S. "Sere, climax, and influent animals with special reference to the transcontinental coniferous forest of North America," *Ecology*, 16:375–402, 1935.

Wiens, J. A. "Avian communities, energetics, and functions in coniferous forest habitats," *Proc. Symp. on Management of Forest and Range Habitats for Nongame Birds*, U.S. Forest Service General Technical Reports, WO-1, 1975.

Desert

Austin, M. *The Land of Little Rain*. New York, Houghton-Mifflin, 1903.

Chew, R. M., and Chew, A. E. "Energy relationships among the mammals of a desert shrub (*Larrea tridentata*) community," *Ecol. Mongr*, 40:1–21, 1970.

Hensley, M. M. "Ecological relations of the breeding bird population of the desert biome of Arizona," *Ecol. Monogr.*, 24:185–207, 1954.

Jaeger, E. C. *The North America Deserts*. Stanford, Cal., Stanford University Press, 1957.

Krutch, J. W. *The Desert Year*. New York, William Sloane, 1952.

Linsdale, J. M. "Environmental responses of vertebrates in the Great Basin," *Am. Midl. Nat.*, 19:1–206, 1938.

Shreve, F. "The desert vegetation of North America," *Bot. Rev.*, 8:195–246, 1942.

Broad Sclerophyll

Balda, R. P. "Foliage use by birds of the oak-juniper woodland and ponderosa pine forest in southeastern Arizona," *Condor*, 71:399–412, 1969.

Cable, D. R. "Range management in the chaparral type and its ecological basis," U.S.D.A. Forest Service, Res Paper RM-155, 1975, pp. 1–30.

Cogswell, H. L. "The California chaparral," *in* O. S. Pettingill, ed., *The Bird-Watcher's America*. New York, McGraw-Hill, 1965.

Hanes, T. L. "Succession after fire in the chaparral of southern California," *Ecol. Monogr.*, 41:27–52, 1971.

Lawrence, G. E. "Ecology of vertebrate animals in relation to chaparral fire in the Sierra Nevada foothills," *Ecology*, 47:278–291, 1966.

Woodbury, A. M. "Distribution of the pigmy conifers in Utah and northeastern Arizona," *Ecology*, 28:113–126, 1947.

Tropical Rain Forest

Beebe, William. *Jungle Days*. New York, G. P. Putnam's Sons, 1925.

Goodnight, C. J., and Goodnight, M. L., "Some observations in a tropical rain forest in Chiapas, Mexico," *Ecology*, 37:139–150, 1956.

Harrison, J. L. "The distribution of feeding habits among animals in a tropical rain forest," *J. Anim. Ecol.*, 31:52–63, 1962.

Janzen, Daniel H. "Herbivores and the number of tree species in tropical forests." *Amer. Nat.*, 104:501–528, 1970.

Odum, Howard T., and Pigeon, R. F., eds. *A Tropical Rain Forest. A Study of Irradiation and Ecology at El Verde, Puerto Rico*. Springfield, Va., National Technical Information Service, 1970.

Richards, P. W. *The Tropical Rain Forest*. New York, Cambridge University Press, 1957.

Tropical Savanna

Beard, J. S. "The savanna vegetation of northern tropical America," *Ecol. Monogr.*, 23:149–215, 1953.

Hemingway, Ernest. *Green Hills of Africa*. New York, Scribners, 1935.

Moreau, R. E. *The Bird Faunas of Africa and its Islands*. New York, Academic Press, 1966.

Petrides, George A. "Big game densities and range carrying capacity in East Africa," *N. A. Wildl. Conf. Trans.*, 21:525–527, 1956.

Aquatic Ecosystems

Cole, G. A. *Textbook of Limnology*. St. Louis, C. V. Mosby, 1975.

Cummins, K. W. "Structure and function of stream ecosystems," *BioScience*, 24:631–641, 1974.

Harvey, H. W. *The Chemistry and Fertility of Sea Water*. Cambridge, Cambridge University Press, 1955.

Hill, M., ed. *The Sea*. New York, John Wiley & Sons, 1962.

Hutchinson, G. E. *A Treatise on Limnology*, Vols. 1–3. New York, John Wiley & Sons, 1957–1975.

———. "Eutrophication," *Am. Sci.*, 61:269–279, 1973.

Hynes, H. B. N. *The Ecology of Running Waters*. Toronto, University of Toronto Press, 1970.

Isaacs, John D. "The nature of oceanic life," *Sci. Am.*, 221 (3):147-162, 1969.

Kajak, Z., and Hillbricht-Ilkowska, A., eds. *Productivity Problems of Freshwaters*. Warsaw, PWN Polish Scientific Publishers, 1970.

Kinne, Otto, ed. *Marine Ecology*, Vol. 1–3. New York, Wiley-Interscience, 1970–1976.

Lauff, G. A., ed. *Estuaries*. Washington, D.C., American Association for the Advancement of Science, Publ. No. 83, 1967.

Likens, G. E., ed. "Nutrients and eutrophication," *Am. Soc. Limnol. Oceanogr. Spec. Symp.* 1, 1972.

Minckley, W. L. "The ecology of a spring stream, Doe Run, Meade County, Kentucky," *Wildl. Monogr.*, 11:5–124, 1963.

Müller, K. "Stream drift as a chronobiological phenomenon in running water ecosystems," *Ann. Rev. Ecol. Syst.*, 5:309–324, 1974.

Prins, R., and Davis, W. J. "The fate of planktonic rotifers in a polluted stream," *Occas. Papers Adams Ctr. Ecol. Studies*, 15:1–14, 1966.

Russell-Hunter, W. E. *Aquatic Productivity*. New York, Macmillan, 1970.

Vallentyne, J. R. "The algal bowl—lakes and man," Dept. of Envir. Misc. Special Publ. 22:1–185, 1974.

Welch, Paul S. *Limnology*. New York, McGraw-Hill, 1952.

Wetzel, R. G. *Limnology*. Philadelphia, W. B. Saunders, 1975.

Whitton, B. A., ed. *River Ecology*. Berkeley, University of California Press, 1975.

Chapter 6

The Practical Ecologist

Probably the single most important message that an environmental scientist has is *caution*: before an action consider its effects, not only now but in the future, and not only at the site of the action but elsewhere in the biosphere. Many practical people would not consider that to be practical advice. To them, "practical" implies action, not inaction; first we try something and then find out what is going to happen, rather than the other way around. But the world has passed the point where this approach is practical; everyday experience no longer equips us to predict the consequences of a new chemical or a new process on the atmosphere or on the genes of the next generation. The old-style practical approach to the environment is similar to that of the man who, after listening to an inspiring sermon, decided to walk upon the water. "How's the water?" his neighbors asked when they found him neck deep in the nearest pond. "Not bad," he replied, "I took two or three steps before I went down."

In its best sense "practical" means putting sound theory into practice. If we are able to do that we may yet get back to shore.

In the next sections we discuss some specific matters of environmental concern to humans and the role of ecology in these concerns.

POLLUTION

It was pointed out in an earlier section that neither pollution nor the scientific study of it are new. What is new, largely a product of the last ten years, is a change in attitude on the part of a great many people. This change has involved the realization, first, that pollution is harmful to humans and the natural systems on which humans depend and, second, that pollution is for the most part preventable. Finally, it has involved the realization that the prevention of pollution should be a cost of doing business, like paying salaries and buying raw materials. When businesses (or governmental agencies) do not pay the cost of preventing pollution, the general public has to pay in medical bills for respiratory ailments, taxes for ever-bigger sewage treatment plants, and in a great many ways that are hard to place a dollar value on, such as having streams with sewage worms instead of trout. Economists call such effects *negative externalities*. When businesses assume the duty of preventing them, as they

247

have increasingly been forced to do, the costs do not thereby disappear; they are paid for instead by higher prices for the company's product and, possibly sometimes, by smaller profits to the company's shareholders. However, if often turns out that, after the immediate changeover costs (such as buying the necessary equipment), pollution control is not very expensive.

TYPES OF POLLUTANTS

Pesticides

Pesticides are materials used to kill pests; *pests* are organisms that interfere with our profit, convenience, or welfare. The definition of pest is, of course, absolutely anthropocentric. Mosquitoes are not pests of the swamps; rather, they are pests to humans living near the swamps. There is no sharp line between what is and what is not a pest but no really exact distinction is necessary, because most pesticides cannot distinguish between a beetle that is or isn't a pest nor, for that matter, between a beetle and a cat.

Early pesticides were the familiar inorganic or organic poisons such as arsenic and nicotine. New organic poisons began to be synthesized beginning in the 1930's and especially in the 1940's. At present these are of three main types: the chlorinated hydrocarbons (such as DDT, DDD, aldrin, dieldrin, heptachlor); the organic phosophorus compounds (such as malathion, parathion, and DDVP); and the carbamates (such as carbaryl and Zectran). So much has been written and spoken about the dangers of pesticides that you would think that everyone who could avoid contact with them would. If you believe this, a visit to your local garden store some Saturday morning will convince you otherwise. Evidently a great many people have not heard of the extreme toxicity of some of these compounds, do not believe what they have heard, or just cannot comprehend it. The last is a distinct possibility. Most people know that arsenic is fairly poisonous and have no desire to have it around, but the fact that parathion is 140 times as poisonous as lead arsenate may be an idea, like the idea of a billion dollars, that is incomprehensible to many people.

Given the potential dangers, surprising amounts of pesticides are used for what can only be regarded as frivolous purposes, such as growing fancy roses and killing harmless but annoying insects. However, most pesticide usage is in agriculture and the control of potentially disease-carrying insects. Although many spokesmen for agriculture–business speak lightly of the dangers of pesticides, most farmers have a realistic idea of the personal hazards of pesticide use. Usually farmers use pesticides because they believe they must, not to raise crops (their fathers grew many of the same crops they grow today without these pesticides) but to compete with other farmers.

The competition is of two forms, the first being yield. If a farmer using pesticides can produce 50 bushels an acre and one not using them can produce only 45, then the pesticide user can sell his crop at a lower price, assuming the pesticide is fairly cheap. If the nonuser has to sell at the same price as the user he may not be able to make a living.

The second kind of competition is for appearance. For instance, pesticides have made possible the routine production of apples of a perfection that, 30 years ago, was rarely found outside the basket of Snow-White's stepmother. The unsprayed apple is often smallish with an occasional worm. These were the sort of apples that everybody ate and made jelly and applesauce from 30 years ago but most people, if given a choice, usually pick the large, shiny, pesticide-produced fruit. Commercial buyers and even governmental agencies now have rules as to "defects" that may make agricultural products raised without pesticides essentially unmarketable.

A good many people have died of pesticide poisoning; nevertheless, the chance of one of us starting up from the breakfast table and falling dead of poisoning from the pesticide residue on our food is slight. Why, then, do environmentalists oppose many of the present practices in pesticide use?

Effects on Nontarget Organisms. Ideally a pesticide should kill only the pest for which it is applied, but most are not very selective. The die-offs of robins from spraying elm trees with DDT (to kill the elm bark beetle that carries Dutch elm disease) became so notorious that they helped to begin the revolt against indiscriminate pesticide use. A group of wildlife biologists from the Illinois Natural History Survey studied the results of an application of dieldrin (3 pounds per acre, spray and granules) for Japanese beetle control. Doubtless some Japanese beetles were killed but the researchers also found dead in the treated areas a great variety of animals, ranging from pheasants, brown thrashers, meadowlarks, ground squirrels, and cottontails, to 14 farm cats.

Persistence and Accumulation. There is, of course, a very good reason why the cats died. They are carnivores, at the top of the food web, where the concentration of pesticides that do not break down rapidly, such as DDT and dieldrin, is greatest. (These problems of persistence and biological magnification are discussed in more detail later in this chapter).

Pesticides do not Stay Where They are Applied. If one farmer sprays his land, his neighbors also get sprayed. Furthermore, persistent pesticides travel long distances in water, air, and by the movement of organisms. DDT is found in the fat of organisms over probably the whole world, including places where no DDT has ever been used; penguins in Antarctica, for example, had DDT levels up to 18 parts per million (ppm) in 1964–65.

Sublethal Effects. Pesticides do not have to kill an organism to harm it. Trout are more susceptible to environmental stresses such as temperature extremes if they are carrying a burden of DDT in their tissues. There is now a large amount of evidence linking DDT to reproductive failure in several kinds of birds, including falcons and pelicans. More than one effect may be involved, but an important one is that DDT and its break-down products cause birds to lay eggs with shells so thin that they cannot hold up under normal incubation.

On the average, you are carrying around in your fat 10 to 20 ppm DDT, 0.1 to 0.3 ppm aldrin and dieldrin (combined), and small amounts of other pesticides. On your food, every day, you probably eat a few tenths or hundreths of micrograms of lindane, dieldrin, malathion, and other pesticides. A microgram is a very small amount, one millionth (0.000001) of a gram, and, in general, the amounts you eat are below the World Health Organization's acceptable daily intake levels. Are you being harmed by these pesticides taken in day after day in small amounts? Are they making you more susceptible to some other diseases, are they af-fecting your behavior, are they causing cancer, are they causing muta-tions? There are studies suggesting that pesticides have these effects and studies suggesting that they do not. The problem is the length of time required for a definite answer. Doesn't the fact that people are living longer today than ever before prove that pesticides, pollution, and all the rest are not really hurting us? No it doesn't, for several reasons. The old folks of today, dying at a ripe age of 70 or 75, were born shortly after the turn of the century; they never met a DDT molecule until they were 50 or so. Will the life span of someone born in 1950 or 1960 also be 70 years of age? It is an interesting question, interesting in a more personal way to someone of 20 than someone of 60.

Synergistic Effects.   As we pointed out in discussing limiting factors in Chapter 2, some environmental factors act more strongly in combina-tion than would be suspected from their effects singly. Malathion is con-sidered a relatively safe pesticide because it is rapidly detoxified by the liver; however, some chemical which interfered with the detoxification function of the liver would increase the danger of malathion poisoning. Just such a chemical is another pesticide, EPN. Malathion was made 50 times more poisonous to dogs when EPN was given at the same time. There are a great many man-made chemicals in use (half a million, ac-cording to the editors of The Ecologist) in pesticides, food additives, hair sprays, medicines, paints, gasoline additives, and many other things. The chance of unsuspected synergistic effects increases as the number of such chemicals increases.

Ecosystem Effects.   The direct effects of a pesticide are rarely con-fined to one species and, furthermore, the overall effects are rarely con-fined just to the species that is harmed directly. The decrease in abun-dance of one species has ramifying effects on other populations at the same and at other trophic levels. Frequently, pesticides are more effective at removing predators of pests than at removing the pests themselves. This occurs because the predators eat a lot of pests, each containing pesticides, and because predators are generally rare relative to their prey, so that a given application of pesticide is likely to kill practically all the predators while leaving a fair number of the pests. If so, the remaining pests can increase, virtually unchecked by the predators. There are now several cases in which applications of pesticides have been followed by an initial reduction in pest density followed by an explosive increase. The European red mite became an important pest in orchards only after the widespread use of chemical pesticides, probably because of their effect on its predators (and possibly competitors).

Resistance to Pesticides.   Most pests have a short generation length and high populations—otherwise they would not be pests. But these two traits favor rapid evolution and consequently rapid development of resistance to pesticides. More than 200 insects are known to have evolved resistance to one or more man-made pesticides in the 30 years since they have been in wide use. The answers to this problem have been, first, to increase the dose. Initially parathion was used against the walnut aphid at a dosage of 0.25 pound per 100 gallons of water, but seven years later the dosage was increased to 1 to 1.5 pounds per 100 gallons. The second step is the development and introduction of a new pesticide. Some insects are reported to have become resistant to chlorinated hydrocarbons, then to organophosphates, and now to the carbamates that have replaced the organophosphates.

Not many insects that are predators or parasitoids of pest species are known to have developed resistance to pesticides, for several reasons. As we have pointed out populations of the predator are usually smaller than those of the prey. Since a new pesticide is generally effective at first in killing a very large proportion of the pest, any resistant predator individuals that survive the pesticide may die of starvation before they can reproduce.

Radioactive Materials

All creatures on this earth are subjected to naturally occurring low levels of high-energy radiation (called *background radiation*) in the form of radioactive material in the earth's crust and cosmic radiation from space. This background radiation may be several times higher on one part of the earth than another. There is no evidence that these natural differences have any ecological effect; there are, however, some statistical correlations that suggest a connection between congenital malformations in humans (such as harelip and clubfoot) and background radiation. Even in areas of high background radiation, however, our exposure is tiny compared with diagnostic X-rays. Two pictures at the dentist's give you as much radiation as you receive from natural sources in the first 15 to 30 years of your life.

Thus it is man-produced radiation with which we are concerned (Table 6–1). Diagnostic X-rays are a health problem but not an ecological

Table 6–1 APPROXIMATE DOSES AND EFFECTS OF DIFFERENT KINDS OF RELEASE OF RADIATION TO OUR NATURAL ENVIRONMENT*

Source	Kinds of Radiation	Duration of Exposure	Dose to Environment	Secondary Activity Induced?	Heat and Blast Effects?	Total Area Involved	Direct Effects?	Incorporated in natural cycles?
Natural (background) radiation	Alpha, beta, gamma	Several billion years	0.1 to 0.5 roentgens per year	No	No	The earth	No	Yes
Medical and occupational	Not normally delivered to man's environment.							
Gamma fields up to 4,000 curies cobalt-60	Gamma	Chronic, several years	Up to several thousand roentgens per hour	No	No	Thousands of acres	Yes	No
Shielded reactor	Mixed gamma-neutron	Intermittent	Up to several times background	Negligible	No	Acres	No	Negligible
Unshielded reactors	Mixed gamma-neutron	Intermittent	Up to 100,000 rads per hour	Yes	No	Hundreds of acres	Yes	Negligible
Reactor effluents	Alpha, beta, gamma	Continuous	Above background	No	No	Hundreds of square miles	No	Yes

Source	Type of radiation	Duration	Intensity			Area		
Waste disposal	Alpha, beta, gamma	Continuous	Slightly above background	No	No	Hundreds of acres	No	No
Accidental explosions	Alpha, beta, gamma	Acute	Up to several thousand rads per hour	No	Yes	Acres	Yes	No
Nuclear detonations	Alpha, beta, gamma	Acute	Up to several million rads per hour	Yes	Yes	Hundreds of square miles	No	No
Fallout from above two sources	Alpha, beta, gamma	Chronic, thousands of years	Up to several times background	No	No	The earth	No	Yes
Nuclear war (projected)	Alpha, beta, gamma	Acute	Up to hundreds of millions rads per hour	Yes	Yes	Thousands of square miles	Yes	No
Fallout from nuclear war (projected)	Alpha, beta, gamma	Chronic, thousands of years	Up to several roentgens per hour	No	No	The earth	Yes	Yes

*From Platt, R., "Ecological effects of ionizing radiation on organisms, communities and ecosystems," *in* Schultz, Vincent, and Klement, A. W., eds., *Radioecology,* New York, Reinhold, 1963, p. 248.

problem. We are concerned here with the release of radioactive materials to the environment, which could occur in two ways: by the testing or actual use of nuclear weapons, or from peaceful applications of nuclear energy, particularly nuclear power plants. In the years since the signing of the nuclear test ban treaty of 1963, in which aboveground testing was outlawed, attention has become focused mainly on radiation release from peaceful applications. This may be too optimistic; both France and Communist China, neither of which signed the nuclear test ban treaty, have recently tested bombs. There is a considerable amount of literature dating from the 1950's and early 1960's on the cataclysmic effects of nuclear warfare (C. L. Comar's "Biological aspects of nuclear weapons" is an example*) which is still current.

Radiation of the sort we are talking about is thought to exert its harmful effect by *ionizing* materials in the cells—that is, by adding or removing electrons. Radioactive materials emit three types of ionizing radiation: (1) *alpha* radiation, which has relatively little penetrating power so that generally a source has to be within the body of the organism to do harm; (2) *beta* radiation, which can penetrate a couple of centimeters of tissue; and (3) *gamma* radiation (similar to X-rays), which has great penetrating power. There are several units of measurement of radiation and radioactive materials; we will mention only the *rad* and the *rem*. Both are measures of the amount of energy absorbed (a rad is an absorbed dose of 100 ergs per gram of tissue and is equivalent to the older term *roentgen*; the rem is similar but is adjusted for the relative biological effects of the different kinds of radiation).

The biological effects of ionizing radiation are of three general sorts:

(1) *Acute*, "radiation sickness" effects resulting from brief exposures to high doses. For humans the LD_{50} is about 500 rads. (LD stands for "lethal dose" and 50 for 50 per cent. An LD_{50}, then, is the dosage at which half those exposed to the radiation live and half die within hours or days.) The 50 per cent surviving generally suffer some ill effects such as cataracts, sterility, or leukemia. Sensitivity to acute exposure varies from one type of organism to another, with mammals among the most sensitive and bacteria among the least.

(2) *Chronic* exposure to low-level radiation may have several effects, of which the best studied is the increased tendency towards cancer.

(3) Related to the chronic effect is the production of *mutations* in sperm and eggs. Geneticists agree that there is no threshold for radiation in its role of increasing mutation. That is, any slight increase in radiation increases mutation rate at least slightly. Most mutations that are detected are harmful, so that this effect of radiation increases the percentage of children bearing harmful mutations.

---

*Am. Sci., 50:339–353, 1962.

The problems involving radioactivity associated with nuclear power plants are of two general sorts: one is the routine release of small amounts of radioactive materials into the air and water, and the other is the possibility of some catastrophe affecting either the plant or the high-level radioactive materials being stored or transported.

Low-level wastes routinely released vary by reactor type, but they include tritium and radioisotopes of iodine, xenon, krypton, strontium, cesium, cobalt, manganese, iron, chromium, molybdenum, and zinc. The release of these materials raises the general level of background radiation. Also troublesome is the possibility that the low-level wastes discharged will become high-level wastes through biological magnification (see p. 257).

High-level wastes must be transported to reprocessing centers, treated, and then stored. Although some countries have dumped sealed containers of these wastes into the oceans, the U.S. has its wastes in temporary storage. In 1977 there were 2500 metric tons of spent fuel stored mostly in swimming-pool-like water basins at power plants, plus 75 million gallons of military wastes stored in underground tanks at three ERDA (Energy, Research and Development Administration) installations. Just where the wastes can be kept is puzzling, because they must be kept for a very long time. Any one batch must be stored for many thousands of years, possibly some few hundred thousand years because of the long half-life of such materials as plutonium.*

The site chosen should then be geologically stable for this length of time, since it would not do to have the material dispersed by an earthquake or a volcano. A geologically stable site may be easier to find than a politically stable one. The United States has just celebrated its Bicentennial, so that its total existence at this point is not one tenth the half-life of plutonium. In fact, the 24,000-year half-life of plutonium would take us back to a time when much of North America was covered with glaciers, when there were still mammoths and saber-toothed cats; and when human culture was still Old Stone Age (in Europe; primitive man had probably not reached North America yet).

Alvin Weinberg, director of Oak Ridge National Laboratory, has written: "We nuclear people have made a Faustian bargain with society. On the one hand, we offer—in the catalytic burner—an inexhaustible source of energy. . . But the price we demand of society for this magical energy source is both a vigilance and a longevity of our social institutions that we are quite unaccustomed to."†

Most opponents of nuclear power believe that such a bargain would be a bad one even if only this generation were affected, and that it is still

---

*The *half-life* of a radioactive substance is the amount of time it takes for half of the radioactive material originally present to decay (transmute).

†From "Social institutions and nuclear energy," *Science*, 177:27–34, 1972.

less desirable to force such a bargain upon future generations. The increased tempo of social change in the past decade, and the possibility of instability as populations continue to rise and resources decline, suggest that our institutions may lack not only the vigilance or longevity but possibly even the technological means of keeping nuclear wastes contained indefinitely.

The safety record of nuclear power plants has been impressively good. Although there have been a few mishaps, none has caused any immediate loss of human life either in the plants themselves or in the surrounding countryside. Nevertheless, some cause for concern remains. In 1975, the Brown's Ferry plant in Alabama came close to a core meltdown when its emergency cooling system malfunctioned. No one is exactly clear what a core meltdown would do but, at a minimum, it would probably produce an explosion (but not a nuclear explosion) that would spread some radioactivity over the countryside in the plant's vicinity.

Even with continued safety in operation there are some other possible dangers. Terrorism could occur at several levels. The possibility exists, and would increase in the future if "breeder" reactors ever become operational, for the theft of the 5 kilograms of plutonium needed to make an atomic bomb, which could then be used to blackmail a city, state, or possibly the Federal governent. However, there really would be no need to go to the trouble of constructing an atomic device. The theft of virtually any fair-sized quantity of concentrated nuclear waste on its way to a reprocessing center and the threat of blowing it up with conventional explosives would be frightening enough.

At this point, of course, "peaceful" uses of atomic energy and warfare begin to merge. India has already set off atomic weapons made from radioactive material supplied to her as fuel for reactors. Senator Mike Gravel has estimated that by 1980 50 countries will have enough plutonium from nuclear power plants and research reactors to construct atomic bombs.

Other Toxic Materials

For some time laws have been in effect that require testing before commercial use of pesticides, drugs, and food additives, and that require safety precautions in the handling of radioactive wastes. These safeguards may not always prove adequate but the legislation does exist. A great many other potentially hazardous chemicals do not fall into any of the categories listed but may enter the environment and do much the same harm as pesticides. Examples are asbestos, which enters the environment from automobile brake bands, talcum powders, and mining wastes; mercury, used as a fungicide, in silent electrical switches, in dental work, and in the paper industry; and PCB's (polychlorinated biphenyls), extremely stable compounds related to DDT and widely used in industry in large electrical insulators, as plasticizers, and in "carbonless" carbon paper. PCB's seem to be at least as harmful as DDT in much the same ways and, furthermore, seem to be as widespread in the bodies of organisms, despite the fact that they were never purposely added to the environment. In September 1976 Congress finally passed legislation (the Toxic Substances Control Act) requiring testing and safeguards in handling these sorts of chemicals. Full implementation of the law will take years.

The potential for trouble that exists with our current, almost uncontrolled use of new toxic materials is illustrated by another compound, PBB (polybrominated biphenyl). Like PCB, PBB is a very stable hydrocarbon with a variety of industrial uses. One of these was in a mixture manufactured by the Michigan Chemical Corporation as a fire retardant and sold under the name "Firemaster." The company also manufactured a grain supplement called "Nutrimaster" to be added to the food of dairy cattle. In the summer of 1973, through a series of events that are slowly coming out during the course of litigation, Farm Bureau Services received a shipment of Firemaster, rather than Nutrimaster, from the Michigan Chemical Corporation and began to mix it into livestock feed which went to 1100 or more farms. Cattle sickened and died but the source of the problem was not identified for several months. By May 1974, PBB had been identified and, when the Michigan Department of Agriculture finally began to investigate, it found PBB in milk and meat throughout much of the state.

The extent of the problem may not be clear for years. One difficulty was the virtual lack of research on PBB's biological effects. The farm families who received the contaminated feed, through drinking the milk from their own herds and eating meat from butchered animals, now have high levels of PBB's in their body. Preliminary medical research has shown that the farm families show a high frequency of certain symptoms, including those of brain damage and a reduced ability to resist infection. To what degree the several million Michigan residents who consumed the contaminated milk and meat suffered the same or other effects may not be known for a long time.

BIOLOGICAL MAGNIFICATION

There is an old saying among sanitary engineers, no longer heard very often, that "dilution is the solution to pollution." Historically, this is one reason why cities and industries tended to locate on rivers, lakes, or the ocean: they could run their wastes into the public waters. Unfortunately, several factors have combined to make "dilution" unsatisfactory as a method of waste disposal. First, there is too much waste being produced; a city or industry of any size just does not have enough water available to it to dump wastes without harming water quality. Second, there are so many people that the solution to one person's pollution may come out of the next person's faucet. If there is only one town on a river, sewage dumped into the water may be decomposed within 20 or 30 miles and the water may be clean again, but if there are towns every 10 miles this recovery may not occur. (One suggested approach to this particular problem, that all users of public water be required to locate their intake pipe downstream from their outlet, has so far met with only slight enthusiasm from most industries and municipalities.)

Another reason why dilution no longer works is slightly more complex. Because of *biological magnification* some materials may be put into the environment dilute and come back concentrated. The concentration occurs in the food chain; the materials that are concentrated are those which the organism takes in but does not readily break down or excrete. The most serious pollutants showing such magnification are some pesticides and other man-made chemicals and radioactive isotopes of certain elements.

Some man-made chemicals such as DDT are extremely persistent—they are not broken down quickly. One reason for this is their novelty; they are new to life and few organisms have evolved the metabolic pathways for decomposing them. DDT does break down eventually, partly through physical and chemical agents and partly through the actions of various organisms that have metabolic pathways to carry on one or two steps somewhere along the line of total breakdown to carbon dioxide, water, and some chlorine-containing compound. Some of the early breakdown products are themselves poisons; DDE apparently is particularly toxic to birds.

DDT (and many of the other chemicals we are concerned about) is soluble in fat but not in water. Consequently, when DDT is in or on the food eaten by a herbivore, it tends to be stored in the fat of the animal, rather than excreted. If the concentration of the pesticide on the plant is some dilute level which we can represent as 1, a particular herbivore may eat 100 plants to stay alive and the concentration in the herbivore may build up to about 100 times (100×) the low level in the plants. In the same way, a carnivore may eat 100 herbivores and the pesticide is magnified to something like 10,000× the concentration at the plant level.

Top carnivores are of course particularly susceptible to being harmed by pesticides as the result of biological magnification. Prime examples are some of the hawks and large seabirds, in which reproductive abnormalities have been shown to be caused by high pesticide levels. Top carnivore fish such as lake trout and Coho salmon in most of the Great Lakes are so contaminated with PCB, DDT, and mercury that women of childbearing age are supposed to avoid eating them, and other persons are advised to eat no more than one-half pound per week.

Radioactive isotopes are concentrated in about the same way. Here the materials that are especially dangerous are mainly the elements essential for some bodily process and that are relatively scarce in the environment. Thus the body tends to retain as much of these materials as possible. Phosphorus, for example, is rare in nature but is part of certain essential cellular compounds (one that has been mentioned is ATP, the P standing for phosphate) and is also concentrated in such places as birds' eggs. One study has shown that radioactive phosphorus in duck and geese flesh was $7.5\times$ the concentration in the water where they fed; the concentration in their eggs was $200,000\times$ the concentration in the water.

Radioactive iodine is another candidate for concentration. Iodine is rare in nature but necessary for producing thyroxin, the hormone of the thyroid gland. Individual organisms concentrate iodine in their thyroids and further concentration may occur up the food chain, through biological magnification.

Some other concentration factors recorded for various radioisotopes have been zinc, $1000\times$ in adult whitefish; iron, $100,000\times$ in algae; strontium, $3900\times$ in muskrat bone; and cesium, $250\times$ in waterfowl muscles.

Neither cesium nor strontium is essential in metabolism but both can be concentrated, cesium being taken up like potassium and strontium like calcium. Both were released in considerable quantities in the early 1960's in the testing of nuclear weapons, and both were found concentrated in human tissue as well as in that of other animals.

POLLUTION IN THE ENVIRONMENT

Water Pollution

Certain populations of humans, especially Europeans and their descendants living elsewhere, have the habit of putting their wastes into water. As discussed in earlier sections, this has the effect of diluting the wastes and, especially in streams, getting them carried out of the vicinity of the producer. Types of pollutants include

sewage and other organic wastes (such as wood fibers from paper mills);

pathogens (mostly bacteria and viruses) from human wastes;

toxic materials (discussed in preceding sections of this chapter) such as mercury, pesticides, and petroleum;

chemicals (usually phosphorus or nitrogen) important in eutrophication (mentioned in the section on limiting factors in Chapter 2 and in the section on lakes and ponds in Chapter 5); and

waste heat (thermal pollution).

Any one source of pollution may, of course, include materials in several of these categories.

For a variety of reasons, relatively large amounts of time and money have gone into studying and attempting to correct the problem of water pollution. These include the fact that water pollution is often readily identifiable by sight or odor and is thus more easily noticed by the public. Second, clean water has obvious economic values (in the price of waterfront property, for example) and public health importance. Third, there seem to be technological solutions to water pollution problems such as building sewer lines and designing large and sophisticated sewage treatment plants. Many other important ecological problems that are more subtle or whose solutions depend on something other than engineering have so far received disproportionately small shares of public attention (and money).

The primary effect of stream pollution by sewage and other organic wastes is in providing large amounts of food for decomposer organisms, mainly bacteria and fungi. The sewage fungus *Sphaerotilus* characteristically appears in organically polluted streams. The respiratory activity of the decomposers removes oxygen from the water, and many characteristic stream animals cannot tolerate the resulting low oxygen levels (Fig. 6–1). Some animals, however, such as sewage worms, may become very abundant.

The term *BOD* (*biochemical oxygen demand*) refers to the amount of oxygen needed to break down (oxidize) organic materials to carbon dioxide, water, and minerals. It is considered that each person generates about 0.17 pound of BOD in a day. Other types of organic pollution can be expressed directly in terms of BOD or converted to PE's (population

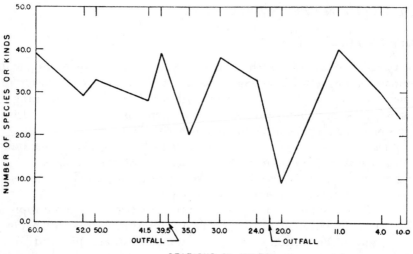

STATIONS IN MILES

Figure 6–1. In this graph, the Mad River, one of Ohio's few trout streams, flows from left to right (the distances are miles above Dayton). At the time of the survey in the 1950's there were two main sources of pollution, Urbana, just below mile 39.5, and Springfield, just below mile 24.0. Most of the pollution was sewage and paper mill waste. The characteristic decline in diversity just downstream from both outfalls is obvious. Species of clean-water forms such as stoneflies and caddis flies dropped out while the pollution-tolerant species such as sewage worms and bloodworms increased. (From A. Gaufin, *Ohio Journal of Science*, 58(4):202, 1958.)

equivalents) by dividing BOD by 0.17. By far the greatest amounts of organic pollution in the U.S. come from the food industry, including raising livestock and processing meat and other foods. The water pollution resulting from a slaughterhouse per ton of slaughtered cattle is 100 to 200 PE's—that is, the same BOD as produced by 100 to 200 people in a day.

The addition of waste heat to water is mainly from electrical generating plants, both conventional and nuclear. Water is taken in to cool the generators and then returned to a lake or stream. Although local modifications of the biota occur in the vicinity of the outfall, it is still unclear how seriously the addition of waste heat is to be regarded. A good-sized power plant located on a small, deep lake might lower the thermocline and extend the summer stagnation period to such a degree that certain types of fish might be harmed. Hot water added to streams tends to eliminate some species of animals and to decrease diversity. In other situations addition of warm water may have somewhat desirable effects—for example, providing open water in winter where waterfowl may feed. The current approach to thermal pollution is to use cooling towers that dissipate the heat to the atmosphere rather than into lakes or streams.

The conclusion of the National Academy of Science that "pollution is often a resource out of place" describes organic and thermal pollution of water very well. The eventual solutions to both problems should be aimed at *using* the organic materials (to enrich soil, for example) and the waste heat, rather than simply trying to dissipate them.

Atmospheric Pollution

Historically, most air pollution has resulted from burning fossil fuels in homes, factories, power plants, and automobiles. From one or all of these sources have come such things as soot, ash, carbon monoxide and carbon dioxide, oxides of sulfur and nitrogen, sulfuric acid, unburned hydrocarbons, formaldehyde, arolein, peroxycetylnitrate, lead, and bromine. Some of these are secondary pollutants resulting from irradiation of the primary pollutants by sunlight, producing *photochemical smog*.

Direct effects of air pollution on human health seem to range from simple irritation of eyes and nostrils through bronchitis and emphysema to lung cancer. During episodes of severe air pollution, death may occur in short order. During the 5-day "killer fog" of early December 1952, 4000 more persons died in London than usually would have died in such a time span. Large-scale conversion from high-sulfur coal to natural gas as a fuel supply in London seems to have made these catastrophes a thing of the past—as long as the natural gas supply holds out.

The same pollutants affecting humans also harm plants and other animals, although there has been little work done on their effects on wild animal populations. Certain types of crops and trees cannot be grown successfully in the vicinity of large cities or heavily travelled highways. Automobile exhaust fumes are known to be harmful to the invertebrate animals living in forest litter. Lichens are sensitive to air pollution, especially to levels of sulfur dioxide (and thus are valuable as indicator species).

Air pollution has ecosystem effects in addition to direct effects on organisms. In some areas, "acid rain" resulting from the formation of sulfuric acid in the atmosphere from sulfur-containing pollutants seems to be acidifying lakes, causing fish production to drop, and to be increasing the nutrient leaching from soils in terrestrial ecosystems. Man is discharging large enough quantities of some elements that pollution has become an important biogeochemical process for the whole biosphere. Combustion of fossil fuels does not add as much carbon to the atmosphere as the respiration of organisms, but it adds more than such sources as volcanoes. Losses from burning fossil fuels and other human activities now add more mercury to the environment than does the weathering of the earth's crust.

As indicated previously (in the section on biogeochemical cycles, Chapter 4), excess carbon dioxide tends to be removed from the atmosphere by the oceans, but the process is slow and some buildup of carbon dioxide in the atmosphere has occurred during the industrial era. This is of concern because of the so-called *"greenhouse effect."* Carbon dioxide is transparent to light but absorbs much of the energy in the infrared, or heat, wavelengths. Consequently, when sunlight hits the earth and warms it (that is, when the light energy is converted to heat energy), the heat radiation from the ground is absorbed by the carbon dioxide of the air, which lets the radiation in the form of light pass through. The result is that the air is warmed and so is the earth, by reradiation from the air. This is a normal process and the temperature of the earth's surface is partly dependent upon it. The fear is that increased levels of carbon dioxide will increase the greenhouse effect and thereby increase the temperatures of the earth. This, in turn, could have many effects—for example, the polar icecaps might melt, causing a rise in sea level and flooding of many coastal areas of the world.

At the same time that man has been putting carbon dioxide into the air, he has also been adding solid particles and droplets that increase the *albedo,* or shininess, of the earth. This should act contrary to the greenhouse effect, reduce the sunlight reaching the earth, and tend to lower temperatures.

Although man seems to have been lucky so far in his thoughtless alterations of the earth, the hope that these two tendencies will turn out to balance one another precisely is, in nontechnical terminology, to expect God to sit in our lap. It is too soon to tell which trend is more important or whether both are minor compared with other influences on world climate. Weather records suggest that the climate has been getting cooler for about the last 30 years after a general warming starting around the end of the nineteenth century.

Worldwide atmospheric pollution can come in less conspicuous packages than the smokestacks of Pittsburgh. Recent studies suggest that the fluorocarbons used as propellants in spray cans are moving to the upper atmosphere, where they are causing ozone to break down to ordinary oxygen molecules. This is significant because (as pointed out in the section *The World's Greatest Reaction* in Chapter 4), the ozone layer, which lies in the zone from about 20 to 55 km—12-35 miles—out from the earth, screens out most of the ultraviolet radiation in sunlight. Decreasing the ozone layer would make it easier to get a suntan but would also raise the frequency of skin cancer and have other potentially harmful effects.

A study committee of the National Academy of Sciences has recommended that the use of fluorocarbons in spray cans be stopped. This is encouraging because it is one of the few cases concerning atmospheric pollution in which the attempted solution has been something other than what environmentalists have referred to as a **"technological fix,"** a situation in which a problem is attacked at the engineering level rather than through a reorientation of attitude or behavior. For example, the solutions up to now for automotive pollution have been to add on "pollution control" devices such as catalytic converters. Technology has been and will continue to be a powerful tool, but purely technological solutions often turn out to be new problems in themselves. Catalytic converters and other pollution control devices, for example, are expensive, may cause engines to run less efficiently—resulting in higher levels of hydrocarbon pollutants and wasting gasoline—and may generate sulfur-containing compounds which under some circumstances may be harmful themselves. Solutions using other than technological approaches (for example, banning private passenger cars from urban areas) have been unpopular but may be beginning to catch on.

FOOD PRODUCTION AND ORGANIC FARMING

There is a wide range of farming practices, from routine heavy usage of herbicides, pesticides, and fertilizers—at present the most common form of commercial agriculture in the U.S.—through various intermediates to totally *"organic" farming*, in which no chemicals are employed for weed or insect control and the only fertilizers used are organic matter, such as manure, and crushed limestone or other rocks. Although the practice of organic gardening and farming has grown in popularity in the past few years, some nutritionists and agricultural experts scoff at organic farming as being hard, unprofitable, and pointless, in that organically grown foods are no more healthful than any other. Advocates of organic farming claim that the current U.S. agricultural system is poisoning us with chemicals, depriving us of, at least, useful trace minerals and possibly other, unknown healthful factors, and leading us down a topsoil-depleted, petroleum-dependent road to ruin.

The facts are far from being well defined. The question of pesticides has already been discussed; the prudent consumer will wish to reduce his intake of chemical residues to as low a level as possible. This is no less true for meat and milk than for fruits and vegetables. Those who consume meat or milk from organically raised livestock, which do not receive food containing synthetic additives, will be safe from the effects of these additives and will also be protected from the effects of mistakes, such as the addition of PBB to livestock feed, discussed in the "Firemaster" episode (p. 257).

It is worth noting that the mere addition of organic matter to soil does not assure its fertility. If neither the original soil nor the organic material added contains much calcium, for example, then calcium will have to be added in some other way. Does it make any difference whether the calcium, phosphate, or nitrogen is supplied "organically" or as manufactured chemicals? It seems very unlikely; the plant absorbs the calcium as an ion from the clay or humus. One calcium ion is much like another.

Are organically grown foods more healthful because of some unknown factors that plants take from the soil or manufacture? Nutritionists correctly state that there is no evidence that this is so. The fact is that there seems to be little evidence at all; the experiments have not been done. It would be poor science to accept the claims of the organic food faddists on this question, but is is equally poor science to say that they are absurd. Science is not based on closed-mindedness.

Of course, environmentalists have suggested that the real food faddists are the food industries. Thirty years ago most of us ate meat, milk, and vegetables that would be considered "organic" today. Except for details, most foods were grown, cooked, and stored just as they had been through human generation after generation. It is primarily since World War II that most of our diets have come to consist of fast-food meals, refined and processed foods, emulsifiers, anticaking agents and shelf-life extenders, spray-dried vegetable oil, sodium sulfite, monosodium glutamate, artificial color, artificial flavor, potassium sorbate, and calcium disodium EDTA (to preserve freshness). This, environmentalists say, is the fad, although it is one in which most of us have been caught up.

The questions of costs and benefits in organic farming are complex. A study by Barry Commoner and his associates at Washington University suggests that organic farms are as profitable as factory farms; however, most economic analyses conclude that each low technology farming practice usually costs more than its high technology alternative.

This conclusion depends strongly on energy costs which, at present, means petroleum costs. Despite the sharp rise in oil prices in the past few

Figure 6–2. Corn production in bushels per acre in the U.S. from 1909 to 1971. This illustrates the increase in productivity that American agriculture was able to achieve starting in the 1940's. (From D. Pimentel, et al., "Food production and the energy crisis," *Science*, 182:445, 1973. Copyright 1973 by the American Association for the Advancement of Science.)

years, fossil fuels still provide a relatively cheap and abundant energy subsidy for man. If energy again becomes expensive, such practices as crop rotation, integrated pest control, greater use of human and animal labor, and less dependence on mechanization may be profitable. Of course this sort of economic analysis is, like most traditional economics, biased for the short term. If the long-term deterioration of agricultural soils, accumulation of pesticides, eutrophication of waters, and disruption of social patterns are included, the price of our current agricultural system may already be more than we would have voted to pay.

There is no question that U.S. agriculture has a remarkable history of productivity if we look no farther than pounds or calories per acre (Fig. 6–2). Much of the increase, however, has come from increasing the energy subsidy to food production, and it appears that the U.S. food system is well along on the curve of diminishing returns (Fig. 6–3).

The *"green revolution"* beginning in the late 1960's was an attempt to increase food production in undeveloped countries by exporting to them a crop production system similar to that of the U.S. New varieties of wheat, rice, and corn were developed that responded better than traditional varieties to fertilization, irrigation, and chemical pest and weed control. Yields were increased substantially, particularly in the period up to 1971. Along with the increase in food production came an increased dependence on energetically expensive technology—for producing such items as the hybrid seeds, fertilizers, pesticides, and tractors—and

Figure 6–3. Farm output plotted against energy subsidies to the U.S. food system from 1920 to 1970. The relationship looks very much like a sigmoid curve that is well past the stage of rapid growth. (From C. E. Steinhart and J. S. Steinhart, *Energy: Sources, Use, and Role in Human Affairs,* North Scituate, Mass., Duxbury Press, 1974.)

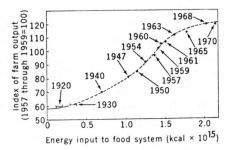

changes in social and political structures. Here, too, the eventual balance of cost and benefit is still uncertain.

ENERGY

Man is uniquely dependent on energy subsidies. A paragraph from H. T. Odum's *Energy, Power, and Society* states the relationship between man and the availability of energy beyond that of his own metabolism in this way:

> Most people think that man has progressed in the modern technological era because his knowledge and ingenuity have no limits—a dangerous partial truth. All progress is due to special power subsidies, and progress evaporates whenever and wherever they are removed. Knowledge and ingenuity are the means for applying power subsidies when they are available, and the development and retention of knowledge are also dependent on power delivery.*

In historic and prehistoric times, man's life was affected by the use of fire and currents, by waterwheels, sailing ships, and windmills, and by animals domesticated as beasts of burden or draft animals. But daily life as we know it is based on abundant, cheap fossil fuel, a development of the nineteenth century.

Current and future energy problems are the result of the same two factors that environmentalists have regularly pointed to as the basis of other environmental problems: an exponentially growing world population and an extravagant technology which, when planned at all, is planned only for the short term.

Given a stable population man's energy needs can be stabilized. Doubtless they can even be decreased with no loss in the quality of life. (Annual per capita energy use in Sweden, for example, is about 60,000 kilowatt-hours vs. 100,000 for the U.S.) But they will remain substantial unless we are willing to forego not just snowmobiles and big cars but also nylon tents and sleeping bags, stereo equipment, and fresh vegetables in the winter.

The solutions to the problem of supplying our energy requirements are easy to see; we must stop population growth, improve the efficiency of our energy usage, and find feasible alternatives to fossil fuels. Only the implementation is difficult.

Efficiency of energy usage includes not only the technical problems of efficiency in converting heat to work but also the whole field of energy conservation. In attempting to conserve energy, it is important to know where efforts would be effective. The breakdown of total fuel energy consumption in Table 6–2 is by end use (electric utility consumption, accordingly, is divided among the uses to which the electricity is put). It is

---

*New York, John Wiley & Sons (Interscience), 1971, p. 27.

Table 6–2 TOTAL FUEL ENERGY CONSUMPTION
IN THE U.S. BY END USE*

End Use	Consumption (trillions of Btu)		Annual Rate of Growth (%)	Percentage of National Total	
	1960	1968		1960	1968
Residential					
Space heating	4,848	6,675	4.1	11.3	11.0
Water heating	1,159	1,736	5.2	2.7	2.9
Cooking	556	637	1.7	1.3	1.1
Clothes drying	93	208	10.6	0.2	0.3
Refrigeration	369	692	8.2	0.9	1.1
Air conditioning	134	427	15.6	0.3	0.7
Other	809	1,241	5.5	1.9	2.1
Total	7,968	11,616	4.8	18.6	19.2
Commercial					
Space heating	3,111	4,182	3.8	7.2	6.9
Water heating	544	653	2.3	1.3	1.1
Cooking	93	139	4.5	0.2	0.2
Refrigeration	534	670	2.9	1.2	1.1
Air conditioning	576	1,113	8.6	1.3	1.8
Feedstock	734	984	3.7	1.7	1.6
Other	145	1,025	28.0	0.3	1.7
Total	5,742	8,766	5.4	13.2	14.4
Industrial					
Process steam	7,646	10,132	3.6	17.8	16.7
Electric drive	3,170	4,794	5.3	7.4	7.9
Electrolytic processes	486	705	4.8	1.1	1.2
Direct heat	5,550	6,929	2.8	12.9	11.5
Feedstock	1,370	2,202	6.1	3.2	3.6
Other	118	198	6.7	0.3	0.3
Total	18,340	24,960	3.9	42.7	41.2
Transportation					
Fuel	10,873	15,038	4.1	25.2	24.9
Raw materials	141	146	0.4	0.3	0.3
Total	11,014	15,184	4.1	25.5	25.2
National total	43,064	60,526	4.3	100.0	100.0

*From Berg, C. A., "Energy conservation through effective utilization," *Science*, 181:131, 1973.
Copyright 1973 by the American Association for the Advancement of Science.

clear that, in homes, ways of decreasing space and water-heating needs will be useful but that foregoing electric toothbrushes, while possibly good for the soul, will not save much energy. Commercial and industrial uses offer possibilities, especially in the use of energetically more efficient buildings and machinery. Air conditioning is growing rapidly in both commercial and residential buildings and, for this reason, is a cause for concern. Large amounts of energy could be saved in our houses, offices, and factories by better design, better insulation, and better placement in the environment.

Transportation offers one of the best targets for conservation. About one fourth of our energy usage goes for transportation, mainly for the private automobile. It has been estimated that the world's automobiles use 12% of the world oil production merely for fuel (and, of course, a large additional percentage for their manufacturing, the building of high-ways, etc.).

A food system that wasted less energy would also be desirable. Agriculture, not treated separately in the table, accounts for 2.6% of the energy used in the U.S.; this includes production of fertilizers and pesticides, planting, and cultivation, but not shipping or processing. As an industry, agriculture ranks third in energy use, after steel and petroleum refining.

It was only about 25 years ago that oil and natural gas came to supply more than half the energy used in the U.S. It is likely that most of us alive today will see them drop from this position, with coal initially becoming more important, as it was in the past. Eventually all fossil fuels must become scarce. As that happens they will become too valuable for the manufacture of plastics and other chemicals to be burnt as fuel. The alternatives to oil, gas, and coal are

> *Hydroelectric power*, including both falling water and tidal installations;
>
> *Solar energy*, including the burning of wood and other recent organic matter, the use of sun and wind for such things as windmills and drying clothes, the use of solar cells to generate electricity, and various "solar heating" applications for houses, hot water heaters, and so forth;
>
> *Geothermal power*, making use of steam or hot water from the earth; and
>
> *Nuclear power*.

NUCLEAR POWER

The Federal government has heavily subsidized the development of a nuclear power industry which, after about 30 years of development, is supplying less than 3% of the energy produced in the U.S. We are concerned here with three kinds of reactors. What might be called at this time the conventional reactor (since there are now about 60 of them in operation) uses the energy from the fission of uranium-235, but there is no long-term future for these reactors because of the scarcity of their fuel. Reserves amount to less than 30 years at projected rates of usage.

The only imminent hope for large quantities of energy from nuclear power was the breeder reactor which, in one version, uses U-235 to convert U-238 (reasonably abundant) into plutonium-239. P-239 is fissionable, as is U-235. Although no commercial breeder reactor has yet been constructed in the U.S. there is no question that they can be built and will work. But there are questions of practicality and safety. Breeder reactors are turning out to be enormously expensive to construct and there seems some doubt that reprocessing centers capable of maintaining the necessary safety and security precautions can operate at an economically practical cost. Of course, there is no purpose in breeding plutonium if it cannot be gotten out of the spent fuel. For these reasons President Carter has taken steps to halt the U.S. breeder reactor program.

The fusion reactor, the third type, would use deuterium, plentiful in seawater, or tritium to produce helium. In this process, which is essentially that used in the sun and the hydrogen bomb, vast amounts of energy would be liberated. Fusion reactors would presumably be less of a pollution problem than the other two types, but this might depend on the specific technical details in commercial application of the process. Any commercial application, however, is far in the future; in fact, there is no assurance that it will ever happen.

SOLAR ENERGY

At the present time, solar energy seems to be the obvious long-term solution to our energy needs. It is abundant and nonpolluting, and it does not add to the heat budget of the earth. A greater diversity of energy sources will probably be used in the future. In contrast to our present dependence on electricity from large power companies, solar houses, windmills, wood-burning stoves, geothermal energy, the burning of solid wastes, and electricity from solar cells (photovoltaic) will be used, depending on the locality and on the individual.

Sunlight is a dilute source of energy—its calories are spread out in time and space—and a great deal of energy will have to be expended just to make it usable for some applications. Accordingly, the future may not be a time of abundant energy. Some environmentalists would worry more about a future in which an abundant energy source had been discovered than they would about a low- or moderate-energy future. They fear that continually growing populations and technology, feeding on abundant energy supplies, would disastrously alter the land, the oceans, and the atmosphere.

POPULATION

Given a limited world, populations cannot continue to grow forever. Even if food, water, energy, pollution level, or some other factor does not become limiting first, present growth rates will fill the land area with people standing next to one another in just under 500 years. One can envision housing everyone in skyscrapers or strange underground cities and adding floating or submarine communities in the lakes and oceans but they would merely postpone matters, even if they were possible; a few hundred years more would be sufficient for the human population to weigh more than the earth. In addition to the logical difficulties of infinite growth in a finite world, our experience with other species shows that the absence of population limitation would be ecologically unique.

If growth cannot continue forever, the appropriate question is when should it stop? Or, a better way of putting the same question, what human population size is optimal?

There is no simple answer but we can begin by realizing that it is a different question from, what is the maximum population that the earth can support? The earth could probably support a higher population living in poverty and distress than in comfortable circumstances, but few would claim that the former are optimal conditions. A second step is in realizing that there is no one answer, no one optimal population, because human preferences are involved in parts of the equation. For example, we might be able to specify an optimal population for a world population living on an almost exclusively vegetarian diet (*almost*, because it may be true that optimal development of children requires fatty acids or other materials found in meat). We might consider safeguarding the land against overuse, preventing pollution of air and water, and take into account the caloric and other nutritional requirements of individuals—and then perhaps we might say that the earth could comfortably support so many billions or millions of people. To allow some fairly substantial part of the world population to include meat in their diet would necessitate a lower population than could be supported—optimally, in purely nutritional terms—on plant material. The freedom to choose a way of life, whether as to diet, religion, or hairstyle, would be held by many persons to be an essential part of an optimal population size.

Although we have been talking on a worldwide scale, three different population sizes are important to humans. The population size of the whole earth is important but so also is the population size of individual nations and of local communities. Determining and achieving optimal populations at each level are separate problems.

For local communities an optimal size is one that is large enough to provide such things as a consumer base for specialized shops and businesses and a social life and clientele for cultural events, and one small enough to avoid the higher crime rates, excessive highway building, destruction of farmland and open space, congestion, and other disadvantages of urban life. Exactly what this optimum may be has not yet been clearly defined. Athens in its golden age had a population of about 40,000 and London in Shakespeare's time was about 180,000. Today in the U.S. it appears that an urban area having 100,000 persons can provide a sufficient audience for most cultural and sporting events, and that per capita taxes for police, education, and welfare go up sharply at a population somewhere between 100,000 and 500,000 (Fig. 6–4).

At some point between one and five million population this trend seems to produce a situation in which a city can no longer support itself.

Figure 6–4. The diseconomies of scale. The per capita costs of police, sanitation and sewage, and highways in Ohio cities are graphed against size of the city. Such services do not just cost more as the size of the city rises; they cost more for each individual, despite the fact that there are more people paying taxes. The biggest city in Ohio is Cleveland with a population of 876 thousand, but the upward trends continue at higher populations. For Los Angeles, with a population of 2.8 million, per capita cost of police protection was $22.39. (Data from Advisory Commission on Intergovernmental Relations, 1968.)

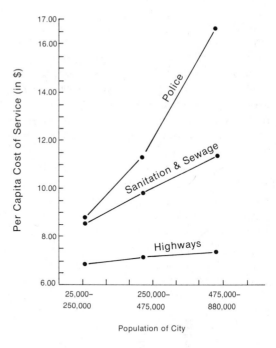

The city's problem then becomes a problem of the rest of the region or of the nation. New York City's financial crisis in 1976 is a case in point. There are two advantages to the nation in having a city as large as New York: (1) certain professions and activities may be so specialized that only cities of a few million people provide a population base big enough to support them; and (2) by concentrating ten million people in a relatively few square miles many of their adverse environmental effects are restricted to a small area. For these reasons a nation may wish to subsidize one or a few very large cities.

Under any circumstances, city and countryside are interdependent in many ways. The city imports food and raw materials for manufacturing and its residents use the open spaces of the farms, forests, streams and lakes in various ways. The countryside is dependent on the city for such things as manufactured goods and services and cash for crops. (There are other sorts of exports, also: the city exports air pollution, after importing the fuels that produce it, and the countryside exports chemical residues in food after importing the pesticides and other compounds.) Here, as previously, the freedom of opportunity to live in rural, small town, or city surroundings is important. So too is the retention of diversity in the landscape so that dwellers in any one of these environments can make use of the unique features of the others.

Controlling local population size has proved particularly difficult in the U.S. for several reasons, despite its importance. Historically, mobility has been an asset for industrial development but such mobility may cause rapid growth of communities through immigration. When local communities have attempted to limit growth, the legal methods used (such as zoning) have sometimes been considered discriminatory by the courts. Probably most important, however, is that most local communities want growth—or to put it more precisely, local leaders want growth. (This matter is dealt with further in the section on land use, p. 277.)

At the national level, current fertility levels of 1.8 children per family would allow a stabilized population of about 250 million (vs. today's 220 million) early in the twenty-first century (ignoring immigration; see p. 76). At that point, a family size of 2.1 children would suffice to keep the population constant. Nationally we seem, then, to be achieving the zero population growth that only a few years ago (say, in 1968, when P. R. Ehrlich's book *The Population Bomb* was first published) seemed unlikely or impossible. Why has the decrease in birth rate occurred? The environmentalists' reasons for controlling population growth—the threat of famine, the role of overpopulation in pollution and land degradation, and all the other unfavorable results of overpopulation—seem not to be the reasons of the persons who have not had the children.

The lowered birth rate seems based on decisions by individuals and couples about the costs and benefits to them of children. Particularly in the 1970's the decisions have been to delay having children to allow time for other activities and to have few or no children—or not to marry—to allow time for careers for women. To be a little less abstract about it, couples are deciding that they want to spend Christmas vacation skiing rather than staying home to overeat and repair toys. Widespread education regarding the world dangers of overpopulation may have had some importance in also providing an altruistic leg for such stands.

Is 220 million Americans optimal? Probably most environmentalists would guess that it is too high. Adding another 30 million to our present population is likely to eat up more land for housing, sewage treatment, highways, forcing still heavier usage of farmland that may already be farmed too intensively to sustain yields without heavy inputs of energy, fertilizers, and pesticides.

Obviously the question is not independent of the size of the world population. Almost no one believes that the earth can support its present 4.3 billion at American levels of affluence. We eat about 3000 calories worth of food a day, including some 65 or 70 grams of animal protein. For the two thirds of the world population in what the food scientist Georg Borgstrom has called the Hungry World, the figures average 2100 calories and 10 grams of animal protein per day. There is about 1 hectare of tilled land per person in the U.S. At this ratio of cropland to population, the world could support about 1.4 billion persons at U.S. levels. Using similar reasoning but different aspects of life and more complicated formulas, several persons have come up with estimates for an optimal world population of around 1 billion.

This leads to one of basically four approaches to the world population problem: it would somehow involve reducing world population by about 75% in order to achieve a high standard of living for everyone. A second approach would be to stabilize world population—or allow it to continue to grow—but to allocate food and other resources evenly so that everyone had a low standard of living. The third approach is to count on the development of abundant cheap energy (as for example by fusion, should it prove technically feasible and inexpensive). If this were to happen, and if the problems of pollution from power plants, transportation, food processing, and so forth could be dealt with, and if unhealthily crowded areas could be reduced without destroying too much of the remainder of the earth, a world population possibly as high as it is now could be supported at a high standard of living.

The fourth approach is that of the present: each sovereign state looks to its own self interests and uses as much of the earth's resources as it can obtain. Although this may seem to be plainly the position of most of the "have" nations, it is not their exclusive property. In recent conferences on population, such as that in 1974 at Bucharest, the usual position of undeveloped countries has been that population size is the business of individual nations rather than a global concern. In these circumstances, those nations which view population size as a worldwide problem can only try to limit their own population and attempt to spread, by appropriate educational measures, a broader realization of the problem.

NATURAL AREAS AND THEIR PRESERVATION

In southwestern Michigan the first settlers claimed lands at the edges of the small prairies. They valued the open land for crops and hay and not for massasaugas or prairie flowers; they valued the oaks for lumber and firewood and acorns for their pigs and not for the orioles that nested there. Next the settlers claimed either grassland or open oak forest. The beech-maple climax forests were called "heavy timber" and were settled last. These forests were thought of mainly as obstacles to raising the crops that the settlers needed to survive, to make a living, and to become successful. The forests were laboriously cleared, at which time the area was said to have "come out of the woods."

Over much of our country forested land is now scarce. Prairie is almost totally gone, having been cultivated continuously since settlement. The oak areas, transformed to farm woodlots, are very different communities from those of primeval times. Bogs and marshes have been drained or filled. Sand dunes have been mined or covered with vacation houses. If we had the chance we might develop our continent more thoughtfully than did our forefathers. We might, but there are no second chances in such matters. The best we can do now is spend our time, energy, and money for preserving the natural landscapes that remain.

A *natural area*, according to the Indiana ecologist A. A. Lindsey and his coworkers, is "any outdoor site that contains an unusual biological, geological or scenic feature or else illustrates common principles of ecology uncommonly well."* No areas on the earth are undisturbed by man, given his use of pesticides, his spread of radioisotopes through the atmosphere, his alteration of water tables, and his many other far-reaching actions. However, most natural areas are sites on which the influence of man has been slight relative to his influence on the rest of the landscape. A beech-maple forest in which no timber was ever cut would be a natural area. So too could be a beech-maple forest in which many trees had been cut 40 years previously. At the present time it might illustrate principles of succession and in the future it might show the workings of a mature ecosystem.

A preserved natural area is one that has been legally dedicated for such a purpose. Natural areas not specifically set aside are lost every year as a landowner interested in preservation dies and his property passes to heirs not sharing his interest, or as university administrations decide that a new fine arts building represents a higher and better use for a piece of land than as an outdoor laboratory for a biology class.

Increasingly, over the past ten years, government agencies, private conservation groups, and a good many ordinary private citizens have begun to acquire and set aside bogs, forests, marshes, and prairies. There are dedicated natural areas in national and state parks, in local nature centers, and as land holdings of many natural history organizations such as the Nature Conservancy and various Audubon Societies.

Specifically, why should natural areas be preserved, set aside permanently as natural areas, when they might well serve as the site of fine arts buildings or houses or highways or sewage treatment plants? To begin with, many of them are irreplaceable, at least in the short term (Table 6–3). This is an idea many persons have trouble with. If a hardware store lies in the path of a new highway the owner can, although he may not want to, build a new store somewhere else. But beech-maple forest destroyed in highway construction is, for practical purposes, permanently lost. To produce a new beech-maple forest starting at bare ground would take hundreds of years of succession. If we get rid of all of the beech-maple forest in our county this year we cannot have some back next year even though we decide we would like some.

Natural areas are places for scientific research on the whole ecosystem, individual plants and animals, and geology and soil science. They are places for education, not only in ecology, zoology, and conservation,

---

*From *Natural Areas in Indiana and Their Preservation,* Indiana Natural Areas Survey, 1969, p. 4.

Table 6–3 ESTIMATED REPLACEMENT TIMES FOR HABITATS IN ILLINOIS*

One aspect of deciding how land should best be used is the "cost" of replacing it, measured in the number of years required to develop an ecosystem once it is destroyed. The right-hand column, "years of replacement time," is the number of years to get back some typical or medium-aged example of the community after disturbance. For example, cutting the large trees in a beech-maple forest produces disturbance, but this may be largely erased in 35 to a few hundred years. If, however, the forest is cut and cleared and the land planted to crops, an additional amount of "successional lead-in time" would have to be added. The cost of replacing each community starting at bare ground, then, is given by adding the two columns together.

Gross Habitats	Years of Successional Lead-in Time	Years of Replacement Time
Bottomland forest		
Oak-gum-cypress	100-150	20–600
Elm-ash-cottonwood by age		
5–29 years	35	5–29
(willow-cottonwood)		
30–59 years	35	30–59
(willow-cottonwood-maple)		
60–99 years	35	60–99
(hackberry-gum)		
100+	135–600	100–500
(hackberry-gum, elm-oak-hickory, and succession to climax)		
Upland forest by age		
10–29 years	25	10–29
(black cherry-elm-hawthorn, elm-persimmon-sassafras)		
30–59 years	50	30–59
(elm-oak-hickory)		
60–99 years	100	60–99
(oak-hickory)		
100+	100+	100–500
(oak-hickory with possible succession to maple-beech)		
Maple-beech	150–200+	35–500+
Aspen	5	5–39
Pine forest by age		
10–39 years	25	10–39
40+	25	40–100+
Shrub areas	3	3–30
Residential habitat		1–100+
Marsh, natural	1,000+	600+
Marsh, man-made	3	3–100+
Prairie	10–15	10–30+
Ungrazed and fallow fields		1–10
Pastures		1–10
Hayfields		1–3
Small-grain fields		1
Row-crop fields		1

*From Graber, J. W., and Graber, R. R., "Environmental evaluations using birds and their habitats." *Ill. Nat. Hist. Surv. Biol.* Note No. 97, p. 1–39, 1976.

but in history and geography. It is difficult to convey the sense of wonder felt by the settlers at their first sight of a prairie if the closest approach to a prairie available is a vacant lot or an oat field.

By providing a place in which species incapable of living on lawns and roadsides can continue to exist, these species with their unique gene pools will be preserved. The potential practical benefits of this should not be dismissed; the possible uses of native organisms—plant, animal, or microbe—for medicines and crops, in biological pest control, and for use in processing of sewage and other wastes have only just begun to be investigated. May apple, for example, is a forest herb that contains the chemical *podophyllotoxin*, which seems useful in treating skin cancer. At the present time the only source of podophyllotoxin is from May apples gathered from the wild, but commercial cultivation of May apple would be desirable. The likelihood of finding strains that could be cultivated and also strains with other desirable traits, such as a high yield of podophyllotoxin, is decreased every time the gene pool of May apple is diminished through the loss of another forest.

There are a variety of other practical uses. The study of stable natural ecosystems may provide insights into how man-influenced ecosystems can be made more stable. The mere existence of natural ecosystems adjacent to crop fields and subdivisions may provide practical benefits in pest control and helping watershed management.

In analyzing the effects of natural areas in the landscape on man, it is difficult to know where to draw the line between practicality and esthetics. Nearly everyone will agree that a forest or an undeveloped lake is valuable as scenery and as a peaceful place to escape from the pressures of civilization. Some would go further and argue that humans are the product of some several hundred thousands of years of evolution, which occurred not in apartment buildings and parking lots but in nature. As adaptable as humans are, it may nevertheless be true, as H. H. Iltis claims, that ". . . for our physical and mental well-being nature in our daily life is an indispensable biological need."*

However, these and other practical values of natural areas may not make very persuasive arguments for their permanent preservation. If salt marshes are preserved only because they process sewage cheaper than man-made tertiary treatment plants, then the marshes must go when sewage technology improves and land prices increase.

Many of the practical values are those of prudence. Some utterly useless marine worm may turn out to cure cancer or some not very pretty ecosystem may be essential for the cycling of sulfur. Economists, who are practical men, do not think much of such values. In their terminology, they discount them. Who knows what the future will bring? It might be nice to have a cure for cancer 30 years from now but by that time cancer

---

*From "A requiem for the prairie," *Prairie Nat.*, 1:51–57, 1969.

may be preventable and a cure may be obsolete. In less practical persons such a view might be called short-sighted.

The practical ecologist, and some others, think of the practical approach as the one which, in the long term, assures the healthful functioning of the earth. To many that approach seems to be the one stated by Aldo Leopold in "The land ethic"*: We are not the boss of the biosphere but just another member of it.

LAND USE

Many environmental problems—population, pollution, natural area preservation—converge in the issue of land use. There are two possible approaches to land use: public planning and what most communities and regions have now.

The land use pattern where you live may be a pleasant patchwork of villages, fields, and forest, or it may be a ghastly mess of smoke-belching factories, ancient and dirty tenements, and misbegotton high-rise apartment projects. Whether it is one of these or something else it probably got that way largely by accident. Even so, it would be a mistake to say that land use planning does not occur. It does, but the planning is mainly in such places as the offices of land developers, paid or unpaid local booster organizations, and road commissions.

Land use decisions—like everything else in the environment—rarely have only a single effect. A decision to build a new factory causes the highway commission to widen the road past the site to four lanes. Motorists begin to use the good, new road as a major artery. No one wants to have their children and dogs next to the speeding cars, the noise, and the fumes, so the zoning board is asked to rezone the neighborhoods along the road for apartment buildings and hamburger stands. These generate more traffic, causing the highway department to put in a six-lane highway which is so convenient as an access into the central city that housing developments spring up around the end of it in the surrounding countryside.

To deal with this sort of ramifying effect and, when necessary to prevent it, regions and communities must have a well thought out plan for land use and evaluate each proposed development on the basis of its eventual as well as its immediate effects.

PHASES IN ENVIRONMENTAL PLANNING

In his textbook, *Ecological Systems and the Environment*, T. C. Foin, Jr., identifies five phases in environmental planning (Fig. 6–5):†

---

*From *A Sand County Almanac*, New York, Oxford University Press, 1949.
†New York, Houghton Mifflin, 1976.

FOREST VALUES

SLOPE

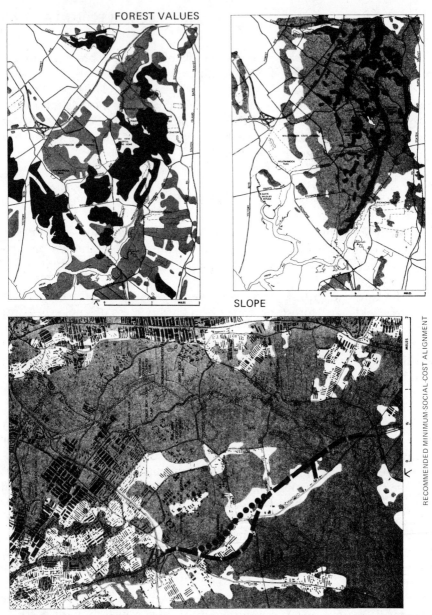

RECOMMENDED MINIMUM SOCIAL-COST ALIGNMENT

Figure 6-5. One of the pioneers and leading practitioners of ecologically based land use planning has been Ian McHarg. The problem that he and his associates dealt with here was smaller than the development of a county or a river basin, but it illustrates the method. A 5-mile section of highway was to be built in the borough of Richmond in New York. Where should it be placed? A series of maps based on a variety of environmental factors was prepared. Two are illustrated: slope and forest values. Areas of steep slope should be avoided for such reasons as expense of construction, susceptibility to erosion, and the fact that these are often scenic areas. Forest and marsh areas of high quality should be avoided also. When these two maps were added to others for historic values, water values, etc., the result was the third map, in which the gray areas are those which should be avoided and the white areas are those in which the highway would result in the least social cost. The recommended route is shown by the dark line. (From I. L. McHarg, *Design with Nature*, Garden City, Natural History Press, 1969.)

1. **An Environmental Resource Inventory.** This should include all environmentally relevant features such as soil types, natural communities, rare and endangered species of plants and animals, human population distribution, air quality, archeological and historical sites, and topography. Aerial photographs and remote sensing data from satellites such as LANDSAT are valuable in the inventory but are not a substitute for actual field work by biologists and other scientists.

2. **Value Inputs and Interpretations.** Ecology and environmental science can carry us only so far in this matter. The recharge areas for ground water can be identified as areas that should not be paved over, the remaining bogs, clean streams, and climax forests can be identified as areas that should be preserved, but in the end a great deal of latitude for judgment is left. It may be a fact that the citizenry is unhealthy without open space, but preferring good health over ill is still a matter of taste. Ideally this phase of planning should include considerations of the total amount of industrialization and commercialization allowable and the total population that would be tolerated. It is important to learn the opinions of the citizens at this stage.

3. **Production of Maps.** Maps of the area are the most straightforward way to assemble the information gathered under Phase 1. By preparing them as transparent overlays (or by computer mapping or other means), the information can be integrated to produce new maps that indicate, for example, the degree of suitability of various regional areas for different uses.

4. **Identification of Development Constraints and the Regional Plan.** The plan is now produced. It should indicate such things as the areas to be preserved, the areas to be retained as agricultural land, the areas on which one residence per acre can be tolerated, and the areas for high density housing.

5. **Implementation.** Implementation of the plan involves such things as the various legal devices for setting it up, cooperation with state and federal agencies and private groups or persons, such as the Nature Conservancy and local philanthropists, and any necessary education of the citizenry.

A few areas have a rationally developed plan for the future which has been developed by some process similar to that just described, but most do not. Few counties, for example, are willing to spend the money to hire an archeologist to compile a map of sites of known and likely archeological importance. Most areas have some type of land use plan or at least a zoning system but up to now such plans have been mostly arbitrary, without the factual basis of a resource inventory and with a value basis having, as its main principle, "growth is good."

Although the past few years have seen several communities decide that they were large enough and would grow no further, most local governments still seem to believe that *any* growth of industry or population is desirable because it "increases the tax base." This ignores the evidence (mentioned under *Population*, p. 270) that beyond some size per capita tax rates increase, despite the fact that there are more people to pay them. Few local governments thoroughly analyze each proposed new business to determine whether the benefits in the form of taxes and payrolls outweigh the costs of such factors as pollution (direct plus indirect, in a form such as auto emissions from employees' cars), loss of open space, provision of utilities, and the tendency of new industry to attract immigration.

Of course, growth is not the only thing valued by local governments. Another important rule seems to be "better us than them." If a new shopping mall is to be located in a region, many local governments are willing to violate their own land use plans so it can be constructed under their jurisdiction rather than in the adjoining city, township, or county. This sort of competition needs to lessened, but whether this can be achieved through regional and state land use plans or by some method of reducing the benefits from winning the competitive game is not clear.

If the U.S. population were 50 million, perhaps even if it were 100 million, land use decisions of this sort could be tolerated. If we did not like what was happening to the neighborhood we could simply move to where it was less crowded and the air was better, just as our ancestors did all through the 1800's. But our population is 220 million, heading towards 250. For most of us there is no longer anywhere to move.

SPACESHIP EARTH

"We travel together, passengers on a little spaceship . . ." wrote Adlai Stevenson and in the dozen years since, the image of "Spaceship Earth" has become a popular one. On short space flights such as those to the moon food can be taken along and wastes stored until the end of the voyage. On longer trips this arrangement would be impossible because of space and weight limitations. The spaceship would have to be complete in itself, producing food and oxygen by the use of photosynthetic plants and recycling wastes as nutrients for plant growth. In the small closed system of the spacecraft the interdependence of the various parts, living and nonliving, is obvious. The spacecraft is, of course, a miniature model of the earth; however, it made a striking image to turn the comparison upside down and think of earth as a spaceship.

The image emphasizes the limitation of space and resources; it is very different from the traditional American idea of moving on to a new frontier further west if we make too big a mess of things here. The image emphasizes interdependencies; we are all here together and earth, air, and water are part of a system to which we ourselves belong. Our continued well-being is dependent on the continued functioning of the whole system.

So the spaceship analogy has its value, but we need to remember that we are not going anyplace; we *live* on "Spaceship Earth." If we find the regimented, claustrophobic conditions of spaceship living a poor sort of home, it is up to us to keep the model right-side up. That is, it is up to us to keep human impact small compared with the earth's capacity to absorb it, so that fields remain fields and oceans remain oceans, rather than becoming "life-support systems." We are travelers together but there is no end of the trip to look forward to when we will get out and stretch, drink clean water, and eat home cooking again.

BIBLIOGRAPHY

Advisory Commission on Intergovernmental Relations. *Urban and Rural America: Policies for Future Growth*. Washington, D.C. U.S. Government Printing Office, 1968.

Armstead, C. H., ed. *Geothermal Energy*. UNESCO, 1973.

Batts, H. L., Jr. "Environmental considerations of the Green Revolution," *Kalamazoo College Rev.* 35:14–16, 1973.

Berg, C. A. "Energy conservation through effective utilization," *Science*, 181:128–138, 1973.

Bertine, K. K., and Goldberg, E. B. "Fossil fuel combustion and the major sedimentary cycle," *Science*, 173:233–235, 1971.

Borgstrom, George. *The Food and People Dilemma*. North Scituate, Mass., Duxbury Press, 1973.

Bresler, Joel B. ed. *Human Ecology: Collected Readings*. Reading, Mass., Addison-Wesley, 1966.

Bromfield, Louis. *Malabar Farm*. New York, Harper & Brothers, 1948.

Carson, Rachel. *Silent Spring*. New York, Houghton-Mifflin, 1962.

Chow, B. G. "The economic issues of the fast breeder reactor program," *Science*, 195:551–556, 1977.

Commoner, Barry. *The Closing Circle–Nature, Man, and Technology*. New York, Alfred A. Knopf, 1971.

Dasmann, R. F. *The Last Horizon*. New York, Macmillan, 1963.

Dochinger, L. S., and Seliga, T. A. "Acid precipitation and the forest ecosystem," *BioScience*, 26:564–565, 1976.

Duffie, J. A., and Beckman, W. A. "Solar heating and cooling," *Science*, 191:143–149, 1976.

The Ecologist, Editors of. *Blueprint for Survival*. New York, Houghton-Mifflin, 1972.

Ehrenfeld, D. W. "The conservation of non-resources," *Am. Sci.*, 64:648–656, 1976.

Ehrlich, Paul R. *The Population Bomb*. New York, Ballantine, 1968.

———, and Ehrlich, Anne H. *Population, Resources, Environment*, 2nd ed. San Francisco, W. H. Freeman, 1972.

Foin, T. C., Jr. *Ecological Systems and the Environment*. New York, Houghton-Mifflin, 1976.

Gibbons, Euell. *Stalking the Wild Asparagus*. New York, David McKay, 1962.

Gibbons, J. W., and Sharitz, R. R. "Thermal alteration of aquatic ecosystems," *Am. Sci.*, 62:660–670, 1974.

Glassman, R., Glassman, U., and Sterne, N. "Have we achieved zero population growth for the wrong reasons?" *J. Envir. Educ.*, 6:12–15, 1975.

Hardin, Garrett. "The tragedy of the commons," *Science*, 162:1243–1248, 1968.

———, ed. *Population, Evolution, and Birth Control*. San Francisco, W. H. Freeman, 1969.

Hubbert, M. K. "The energy resources of the earth," *Sci. Am.* 224(3):60–87, 1971.

Hynes, H. B. N. *The Biology of Polluted Waters*. Liverpool, University of Liverpool Press, 1960.

Johnson, W. A., Stoltzfus, V., and Craumer, P. "Energy conservation in Amish agriculture," *Science*, 198:373–378, 1977.

Klein, L. *River Pollution*. Woburn, Mass., Butterworths, 1966.

Lamm, R. D. "Local growth: focus of a changing American value," *Equilibrium*, 1(1):4–8, 1973.

Leopold, Aldo. *A Sand County Almanac and Sketches Here and There.* New York, Oxford University Press, 1949.

Lindsey, A. A., Schmelz, D. V., and Nichols, S. A. *Natural Areas in Indiana and Their Preservation.* Lafayette, Indiana Natural Areas Survey, 1969.

Lockeretz, W., Klepper, R., Commoner, Barry, Gertler, Michael, Fast, S., O'Leary, D., and Blobaum, R. "A comparison of the production, economic returns, and energy intensiveness of corn belt farms that do and do not use inorganic fertilizers and pesticides." *Ctr. Biol. Nat. Syst., Rep. No. CBNS-AE-4,* 1975.

Maddox, J. *The Doomsday Syndrome.* New York, McGraw-Hill, 1973.

Marsh, George Perkins. *Man and Nature, or Physical Geography as Modified by Human Action.* New York, Scribners, 1864. Reprinted by Harvard University Press, Cambridge, 1965 (D. Lowenthal, ed.).

Mayer, J. "The dimensions of human hunger," *Sci. Am.* 235(3):50–73, 1976.

McCaull, J. "Windmills," *Environment,* 15(1):6–17, 1973.

McHarg, Ian L. *Design with Nature.* Garden City, N.Y., Natural History Press, 1969.

Meadows, D. H., Meadows, D. L., Randers, J., and Behrens, W. W., III. *The Limits to Growth.* Washington, D.C., Potomac Associates, 1972.

Morris, D. N. *Future Energy Demand and its Effect on the Environment.* Rand Corporation, 1972.

Newman, O. *Defensible Space.* New York, Macmillan, 1972.

Odum, Howard T. *Environment, Power, and Society.* New York, John Wiley & Sons (Interscience), 1971.

———, and Odum, Elizabeth C. *Energy Basis for Man and Nature.* New York, McGraw-Hill, 1976.

Owen, O. S. *Natural Resource Conservation: An Ecological Approach,* 2nd ed. New York, Macmillan, 1975.

Pimentel, David, et al. "Food production and the energy crisis," *Science,* 182:443–449, 1973.

———, et al. "Land degradation: effects on food and energy resources," *Science,* 194:149–155, 1976.

Pollard, W. G. "The long-range prospects for solar-derived fuels," *Am. Sci.,* 64:509–513, 1976.

Putnam, P. C. *Energy in the Future.* New York, Van Nostrand, 1953.

Resources for the Future. *U.S. Energy Policies: An Agenda for Research.* Baltimore, Johns Hopkins, 1968.

ReVelle, Charles S., and ReVelle, Penelope L., *Sourcebook on the Environment.* New York, Houghton-Mifflin, 1974.

Romancier, R. M. "Natural area programs," *J. Forestry,* 72:37–42, 1974.

Schipper, L., and Lichtenberg, A. J. "Efficient energy use and well-being: the Swedish example," *Science,* 194:1001–1013, 1976.

Schumacher, E. F. *Small is Beautiful.* New York, Harper & Row, 1973.

Shepard, Paul, and McKinley, Daniel, eds. *The Subversive Science: Essays Toward an Ecology of Man.* New York, Houghton-Mifflin, 1969.

Singer, S. F., ed. *Is There An Optimal Level of Population?* New York, McGraw-Hill, 1971.

Spanides, A. G., and Hatzikidis, A. D. eds. *Solar and Aeolian Energy.* New York, Plenum, 1964.

Spilhaus, Athelstan, ed. *Waste Management and Control.* Washington, D.C., National Academy of Science, 1966.

Stadtfeld, C. K. "Cheap chemicals and dumb luck," *Audubon,* 78(1):110–118, 1976.

Steinhart, J. S., and Steinhart, C. E. "Energy use in the U.S. food system," *Science,* 184:307–316, 1974.

Teitelbaum, M. S. "Relevance of demographic transition theory for developing countries," *Science,* 188:420–425, 1975.

Thomas, William L., Jr., ed. *Man's Role in Changing the Face of the Earth.* Chicago, University of Chicago Press, 1956.

Wade, N. "Control of toxic substances: an idea whose time has nearly come," *Science,* 191:541–544, 1976.

Wagner, Richard H. *Environment and Man,* 2nd ed. New York, Norton, 1974.

Warren, Charles E. *Biology and Water Pollution Control.* Philadelphia, W. B. Saunders, 1971.

GLOSSARY

Many words not included here are defined explicitly or by context in the body of the book. For words not given here, consult the index. Several terms considered unnecessary for this book but which may be encountered in other readings in ecology are defined here. Two useful sources for ecological terms, although both are now somewhat dated, are J. R. Carpenter's *An Ecological Glossary* (Norman, Okla., University of Oklahoma Press, 1938, reprinted in 1956 by Hafner) and H. C. Hanson's *Dictionary of Ecology* (New York, Philosophical Library, 1962). For general biological terminology, E. B. Steen's *Dictionary of Biology* (New York, Barnes and Noble, 1972) and *The American Heritage Dictionary of the English Language* (New York, Houghton Mifflin, 1969) are useful.

Abiotic. Nonliving.
Acclimation. Phenotypic adjustment (especially change in tolerance limits) to changed conditions.
Allelopathy. Chemical inhibition of one organism by another.
Anaerobic. Without oxygen.
Annual. A plant with a life span of a year.
Aspection. Seasonal change.
Association. (1) A community recognized on the basis of a characteristic combination of species; (2) a subdivision of a formation.
Autecology. The ecology of the individual organism.

Biocenose. Community.
Biogeochemical cycle. A circular system through which some element is transferred between the biotic and abiotic parts of the biosphere.
Biomass. The weight of living material (of a population, trophic level, etc.)
Biome. A community of geographical extent characterized by a distinctive landscape based on the life forms of the climax dominants.
Biota. The organisms of an area.
Biotic. Pertaining to life.
Biotic potential. The inherent capacity of a population to increase in numbers.

B.O.D. Biochemical oxygen demand. Oxygen required to break down organic material and to oxidize reduced chemicals (in water or sewage).

Bog. A type of wetland ecosystem in which organic matter accumulates as peat.

Carrying capacity. The maximum number of a species that can be supported indefinitely by a certain area.

Circadian rhythm. A cycle of activity having an approximately 24-hour period.

Climax community. A community that is the stable end product of succession.

Coaction. The effect of one organism on another.

Coevolution. The simultaneous evolution of two or more coacting species.

Community. A group of organisms occupying a particular area; the connotation is of a coacting system.

Competition. A combined demand in excess of the immediate supply.

Conservation. Preservation and wise use of natural resources.

Continuum. A gradient of change in species composition related to habitat.

Culture. Socially transmitted behavior patterns, beliefs, language, institutions, etc., characteristic of a human population.

Demography. The study of characteristics of populations (especially humans).

Density. Numbers per unit area.

Detritus. Dead particulate organic matter.

Diel. Daily.

Dispersal. The movement of individuals from the homesite.

Diversity. Variety; usually refers to number of species in a community and often includes some measure of their relative abundance.

DOM. Dissolved organic matter.

Dominance. Control exerted on the character and composition of an ecosystem by an organism.

Dominance hierarchy. Social organization in which high-ranking individuals have precedence in the use of resources.

Ecological amplitude. Range of tolerance.

Ecological efficiency. The percentage of energy taken in (as food) by one trophic level that is passed on as food to the next higher trophic level. Several other ratios of energy transfer may also be calculated.

Ecological equivalents. Unrelated species playing the same ecological role (occupying the same niche) in different geographical areas.

Ecological indicator. A species or community (or their response) that is a measure of environmental conditions.

Ecological niche. The role of a species in an ecosystem.

Ecology. The study of the relationships of organisms to their environment and to one another.

Ecosystem. The community plus its habitat; the connotation is of an interacting system.

Ecotone. A transition zone between two ecosystems.

Ecotype. A genetically differentiated population within a species, the differences having ecological significance.

Edaphic. Pertaining to soil.

Environment. The surroundings of an organism.

Environmental impact statement (EIS). An examination of the potential environmental effects of a proposed action including a consideration of alternative courses of action, required for Federal projects by the National Environmental Policy Act.

Environmental resistance. The effects of crowding that lower population growth ·below that potentially possible.

Estuary. An arm of the sea where fresh and salt water mix.

Ethology. The study of the function, causation, and evolution of behavior.

Eutrophication. The increase of nutrients in lakes either naturally or artificially by pollution.

Evapotranspiration. Loss of water from a land surface through both transpiration from plants and evaporation.

Evolution. A change in gene frequencies with time; the adaptive modification of organisms through successive generations.

Fauna. The animals of an area.

Fecundity. The potential level of reproduction in a population. In human populations, for example, slightly more than one birth per reproductive-age female per year.

Fertility. The realized level of reproduction in a population; the actual number of live births.

Flora. The species of plants in an area.

Food chain. A sequence of organisms, each of which feeds on the preceding.

Food web. A trophic system composed of interconnected food chains.

Forb. A herb that is not a grass, sedge, or rush.

Forest edge. The vegetational structure characteristic of the edge of a forest, a type of savanna with interspersed trees, thicket, and herbaceous areas.

Formation. One of the great subdivisions of the earth's vegetation (deciduous forest, tundra, etc.) having a distinctive physiognomy based on the life forms of the climax dominants.

Fugitive species. Species characteristic of temporary habitats.

Geographic range. The area of the earth's surface occupied by all the populations of a species.

Guild. A group of species that share a resource (have related niches) in a community.

Habitat. The specific set of environmental conditions under which an individual, species, or community exists. Sometimes restricted to conditions of the physical environment; sometimes restricted to individuals or species (in which case the habitat of a community is termed a *biotope*).

Home range. The area routinely traversed by an animal.

Hydric. Wet.

Life form. The characteristic adult growth form of an organism (especially plants); usually does not refer to the Raunkiaer life-form system unless so specified.

Life table. Vital statistics, by age, of a population.

Limiting factor. The factor which, by its relative scarcity or unfavorability, limits a process or the numbers or range of an organism.

Logistic curve. An S-shaped population growth curve in which growth rate begins to decline about halfway up.

Mesic. Having or characteristic of a medium moisture content.

Monoculture. A single crop grown over a large area.

Mycorrhiza. A symbiotic association of vascular plant roots and fungus, important in nutrient uptake by the plant.

Natural area. A site (usually with minimal recent human disturbance) that contains an unusual biological or other natural feature, or that illustrates ecological principles unusually well.

Natural selection. The preservation from generation to generation of favorable individual differences; differential perpetuation of genes in successive generations caused by different degrees of adaptedness to the environment; survival of the fittest.

Old field. Abandoned cropland.

Open space. Land that is sparsely occupied by human construction; natural areas, parks, and farmland.

Organic farming. Agriculture in which synthetic chemicals are avoided and fertilizers in the form of composted organic matter and crushed rocks are employed.

Paleoecology. The study of the interactions of organisms and environment in the geologic past.

Perennial. A plant with a life span of more than two years.

Permafrost. Permanently frozen soil.

Phenology. The study of seasonal change.

Photoperiod. Day length.

Pioneer community. The initial community in succession.

Pollution. The unfavorable modification of the environment by human activity.

Population. A set of organisms belonging to the same species and occupying a particular area at the same time.

Production. Energy storage; increase in organic matter.

Productivity. Rate of energy storage or increase in organic matter.

Quadrat. A sample plot (properly, rectangular)

Reaction. The effect of organisms on their physical environment.

Sample. A portion chosen from a population. A *random sample* is one in which every individual (organism, point, plot) has an equal and independent chance of being chosen.

Sere. The whole sequence of communities leading to the climax. *Seral stage,* any community in a sere; a successional community.

Shade tolerance. The ability of a plant to survive and grow in shade.

Species. A group of populations reproductively isolated from other such groups.

Standing crop. Biomass (rarely, numbers) present at a given time.

Stratum. A layer of an ecosystem.

Succession. The change in numbers and kinds of organisms leading to a stable (climax) community. Replacement of communities, one by another, on an area.

Symbiosis. Any intimate coexistence such as parasitism or mutualism.

Synecology. Community ecology, especially the habitat relations of communities.

Synusia. Plants of the same life form, often forming a layer.

Territory. A defended area.

Threshold. The intensity (of a stimulus) below which no response is produced.

Tolerance, Limits of. The lower and upper survivable levels of some physical factor for an organism or population.

Tolerance, Range of. The set of conditions for some physical factor within which an organism or a population can survive.

Trophic. Feeding; hence, related to energy transfer.

Vegetation. The plant cover of an area.

Wilderness. Land unaffected, or not evidently changed, by human presence.

Xeric. Dry.

Yield. Organic production used by men.

ZPG. Zero population growth; a situation in which birth and immigration rates are balanced with death (and emigration) rates, so that a population neither increases nor decreases with time.

INDEX

Page numbers in *italics* indicate illustrations.
Page numbers followed by (t) indicate tables.